Eugene Chaikovskaya

Smart Grid Technologies in Electric Systems for Wind and Solar Energy

www.novapublishers.com

NOTICE TO THE READER

Library of Congress Cataloging-in-Publication Data

ISBN: 979-8-89113-446-1

Published by Nova Science Publishers, Inc. † New York

Contents

Abstract

Integrated Smart Grid systems for the coordination of electric power production and consumption based on forecasting the change in parameters of technological processes are developed. Advance decisions to support the operation of electrical systems are based on the mathematical description of the architecture of technological systems, the methodology of mathematical description of the dynamics of energy systems, the cause-effect graph method, and the mathematical description for operational maintenance of Smart Grid electric systems. For example, the developed integrated Smart Grid system to support the operation of a wind-solar electric system is based on predicting a change in the capacity of a rechargeable battery by measuring voltages at the hybrid charge controller input, the inverter output and the current frequency. The advance decisions made to maintain the capacity of a rechargeable battery related to a change in the capacity of a thermoelectric battery are based on establishing the ratio of the voltage at the hybrid charge controller input to the voltage at the inverter output, which are measured. A change in the rotational speed of the electric motor in the circulating pump optimises heated water consumption and temperature, thus reducing the charging duration by up to 30%. The basis for the proposed technological system is a dynamic subsystem that includes the following components: a wind-energy installation, a photoelectrical electrical module, a hybrid charge controller, an inverter, an array of rechargeable batteries, and a thermoelectric battery. A functional assessment has been derived for a change in the capacity of a rechargeable battery, the rotational speed of the electric motor in the circulating pump, and local water consumption related to changes in the temperature of local water within the range of 30–70°C. Defining the resulting functional information on forecasting a change in the capacity of a rechargeable battery enables making the following advance decisions: on changing the rotational speed of the electric motor in the circulating pump and on changing the local water consumption. The capacity of a rechargeable battery is maintained by adjusting energy generation and consumption. For example, the developed integrated Smart Grid system matches electric power production and consumption based on a prediction of changes in the battery capacity. Advance decisions on the change in power transmission capacity enable regulating the voltage within the distribution system by maintaining the power factor of the photoelectric charging station. Voltages at the hybrid inverter input and within the distribution system are measured to assess their ratio. A functional estimate of changes in the battery capacity and power factor of the photoelectric charging station is obtained. Maintenance of voltage within the distribution system is performed based on the resulting operational data to estimate a change in the battery capacity. Advance decision-making enables increasing the power factor of the photoelectric charging station by up to 40% by matching the electric power production and consumption. The operational maintenance of the photoelectric charging station using the developed Smart Grid technology enables the prevention of peak load on the power system through a 20% reduction of power consumption in the network.

Keywords: Smart Grid technologies, wind and solar energy, forecasting of change of parameters, operational maintenance

Introduction

The book includes seven chapters covering integrated Smart Grid systems for the coordination of production and consumption of electricity in wind and solar electric systems using forecasting of changes in technological process parameters that are being developed. Advance decisions to support the functioning of electrical systems are based on the mathematical description of the technological system architecture, the methodology of mathematical description of dynamics in energy systems, the cause-and-effect graph method, and the mathematical description of operational maintenance of Smart Grid wind and solar electric systems.

Chapter 1 – The mathematical substantiation of operational maintenance of Smart Grid wind and solar electric systems based on the architecture and mathematical description of the architecture of wind and solar electric systems, methodology of the mathematical description of power systems dynamics, and the cause-effect graph method is proposed. The mathematical substantiation is based on the systematic approach to complex mathematical and logical modelling of wind and solar electric systems as dynamic systems. Decision-making to support the operation of energy systems is based on the mathematical description of the architecture of wind and solar electric systems, the methodology of mathematical description of the dynamics of energy systems, and the cause-effect graph method. Coordination of energy production and consumption is based on forecasting changes in the parameters of technological processes. Real wind and solar electric systems are dynamic systems, the mathematical models of which reflect the properties of the transformation of influences, that is, their dynamic properties. The operational maintenance of wind and solar electric systems takes place in the composition of such wind and solar electric systems, which are based on dynamic systems. The design of wind and solar systems is based upon integrated dynamic subsystems for evaluating changes in both production and consumption of energy. The design of technological systems is represented as an organization of complex systems, expanded by building dynamic subsystems of nodes that predict the components of the technological processes as its basis. Considering the system-structural and mathematical substantiation of the architecture of wind and solar electric systems, the relation category can be viewed as organized interactions not only within the elements of the technological systems but also within the elements of the integrated dynamic subsystems, which enable performing operability control and identifying the state of energy systems based on the developed cause-effect graph method. The final assessment, obtained using logical relations in dynamic subsystems, enables the implementation of new properties into energy systems through the appropriate decision-making and the identification of new operating conditions. Moreover, the relations between the dynamic subsystems and the blocks within their composition enable, based on the assessment of the state of the parameters diagnosed within these blocks, confirming the new operational conditions of the energy systems.

Chapter 2 – Based on mathematical and logical modelling, a technological support system for changing the battery capacity created on the prediction of voltage variation through measuring the electrolyte temperature within the battery volume was developed in the composition of a technological system for battery operation. The developed technology enables the following: controlling the operational capacity of the accumulator battery in order to obtain a functional assessment of change in the total charge and discharge voltage; obtaining an integrated reference estimation of change in the charge and discharge voltage; developing an integrated system for assessing a change in the battery voltage, thus enabling the maintenance of the accumulator battery capacity by measuring electrolyte temperature at the battery input. The limit of electrolyte temperature change to 35°C while charging with direct current and limited voltage change for further charging and discharging was established with a change in the consumption of electric energy. The use of an integrated system for the estimation of voltage change obtained based on the alignment between electrochemical and diffusion processes of discharging and charging enables making timely decisions on recharging to prevent overcharging and unacceptable discharge. Coordination of the electrochemical and diffusion processes that accompany the charging and discharging of the battery enables a reduction in energy production costs, for example, when operating a wind power plant with a capacity of 10 kW, and the payback period by up to 25% thanks to the reduction of charging duration and the prevention of gas formation.

Chapter 3 – On the basis of mathematical and logical modelling as part of the technological system for the operation of thermal electric accumulators, a system of technological support for changing the capacity of accumulators was developed based on predicting changes in the temperature of heated water within the accumulator volume by measuring the water temperature at the thermoelectric accumulator inlet and its outlet. The developed technology enables controlling the operation of a thermal electric accumulator to obtain a functional assessment of the change in the heated water temperature. The use of an integrated system for estimating the change in the heated water temperature within the battery volume, obtained by matching the heat and mass transfer in discharging and charging processes, thus enabling timely decisions on changing the thermal electric accumulator power based on the change in local water flow rate, reducing the charging time by up to 30%. For example, this technology is used in the developed integrated Smart Grid system to support the operation of a wind-solar power plant by predicting changes in battery capacity (Chapter 6). Changes in the rotation speed of the circulation pump electric motor are provided in the event of a change in the heated water flow rate and temperature, thus reducing the charging duration by up to 30%. The storage battery and the thermoelectric accumulator, as part of the solar power system network (Chapter 4), acquire the additional status of voltage regulators within the distribution system. Advance decisions are made to change the capacitance of a thermoelectric accumulator during the redistribution of the accumulated electrical energy. Changing the level of transmission of electrical energy to the grid enables maintaining the voltage within the distribution system by adjusting the power factor in the solar power plant grid. The peak load on the power system is prevented, which reduces the consumption of electricity from the network by up to 14%.

Chapter 4 – It is proposed to maintain the voltage within the distribution system by forecasting the change in the capacity of the storage battery to make anticipatory decisions on the change in the power of the thermoelectric accumulator to maintain of the power factor of the solar electric system network. There is an assessment of the change in the ratio of voltages at the frequency converter output and within the distribution system, which are measured by measuring the voltage at the hybrid inverter input. A comprehensive system has been developed to support the operation of a solar power plant network based on predicting changes in battery capacity and power factor. As part of the solar power system network, the accumulator battery and the thermoelectric accumulator acquire the additional status of voltage regulators within the distribution system. Advance decisions are made to change the capacity of a thermoelectric accumulator in the redistribution of accumulated electrical energy. Changing the level of transmission of electrical energy to the grid enables maintaining the voltage within the distribution system by maintaining the power factor in the solar power plant grid. The voltages at the hybrid inverter input, the frequency converter output, and within the distribution network are measured continuously. The change in the ratio of voltages at the frequency converter output and within the distribution network is estimated. Peak load on the power system is prevented, thus reducing power consumption from the grid by up to 14%.

Chapter 5 – The Integrated Smart Grid System for harmonization of electric power production and consumption based on a prediction of changes in the battery capacity is developed. The integrated dynamic subsystem of the photoelectric charging station includes the following components: mains, photoelectric solar panels, a hybrid inverter, rechargeable batteries, a two-way Smart Meter and a charger. Advanced decisions on the change in power transmission capacity Enabled regulating the voltage within the distribution system by maintaining the power factor of the photoelectric charging station. Voltages at the hybrid inverter input and within the distribution system were measured to assess their ratio. Comprehensive mathematical and logical modelling of the photoelectric charging station was performed based on the mathematical substantiation of architecture and operational maintenance. A dynamic subsystem including such components as mains, a photoelectric module, a hybrid inverter, batteries, a two-way Smart Meter and a charger formed the basis of the proposed technological system. Time constants and coefficients of dynamic mathematical models were determined in terms of the estimation of changes in the battery capacity and power factor of the photoelectric charging station. A functional estimate of changes in the battery capacity and power factor of the photoelectric charging station was obtained. Maintenance of voltage within the distribution system was realized based on the resulting operational data to estimate a change in the battery capacity. Advance decision-making has made it possible to raise the power factor of the photoelectric charging station by up to 40% due to matching the electric power production and consumption. Operational maintenance of the photoelectric charging station using the developed Smart Grid technology has enabled the prevention of peak load on the power system due to a 20% reduction in power consumption from the network.

Chapter 6 – The developed integrated Smart Grid system for supporting the operation of a wind-solar electric system is based on predicting a change in the capacity of a rechargeable battery by measuring the voltage at the hybrid charge controller input, the voltage at the inverter output and the current frequency. Making advance decisions to support the capacity of a rechargeable battery related to a change in the capacity of a thermoelectric battery is based on establishing the ratio of voltage measured at the hybrid charge controller input to the voltage at the inverter output. A change in the rotational speed of the electric motor of the circulating pump has been ensured in terms of changes in consumption and the temperature of heated water by reducing charge duration by up to 30%. The basis for the proposed technological system is a dynamic subsystem that includes the following components: a wind-energy installation, a photoelectrical electrical module, a hybrid charge controller, an inverter, an array of rechargeable batteries, and a thermoelectric battery. A functional assessment has been derived for a change in the capacity of a rechargeable battery, the rotational speed of the electric motor of the circulating pump, and local water consumption related to a change in the local water temperature in the range of 30–70°C. Defining the resulting functional information on forecasting a change in the capacity of a rechargeable battery enables making the following preliminary decisions: on changing the rotational speed of the electric motor of the circulating pump and the consumption of local water. Maintaining the capacity of a rechargeable battery is carried out based on adjusting the generation and consumption of energy.

Chapter 7 – Integrated Smart Grid Systems for harmonization of production and consumption of electric power of the heat pump power supply and hot water power supply are being developed. The integrated dynamic subsystem of the wind-solar electric system includes the following components: the electric network, wind turbine, photovoltaic solar panels, hybrid solar collectors, grid inverter, heat pump, two-section storage tank, upper section for hot water supply, lower section a low-grade energy source, and a frequency converter. Integrated systems based on predicting changes in the power factor and local water temperature by measuring voltage from hybrid solar collectors at the grid inverter input, voltage at the frequency converter output, and current frequency. The adoption of advance decisions to maintain local water temperature by changing the power of the electric motor in the heat pump compressors and the electric motor of the circulation pump is based on establishing the ratio of the voltages at the grid inverter input and the frequency converter output. The power factor of the wind-solar electric system is maintained. The complex mathematical and logical modelling of the wind-solar electric system, based on the mathematical substantiation of the architecture of the wind-solar electric system and mathematical substantiation of the operational maintenance of the wind-solar electric system is performed. Time constants and coefficients of the mathematical models of dynamics regarding the estimation of a change in the power factor of the system, and the local water temperature are predicted by measuring voltage from hybrid solar collectors at the grid inverter input, voltage at the frequency converter output, and current frequency. Functional estimation of a change in the power factor of the wind-solar electric system is within the range of 58–98%, and local water temperature is within the range of 30–55°C. Determining final functional information provides an opportunity to make advance decisions on a change in the operation of the electric motor in the heat pump compressor and the electric motor in the circulation pump to prevent the peak load on the power system and maintain voltage when the heat pump power supply and hot water supply are connected.

Chapter 1

Methodological and Mathematical Substantiation of Smart Grid Technologies for Maintaining the Operation of Wind and Solar Electric Systems

Abstract

This chapter proposes a mathematical substantiation of operational maintenance of smart grid wind and solar electric systems based on the architecture and mathematical description of the architecture of the wind and solar electric systems using the methodology for the mathematical description of the dynamics of power systems and the cause-effect graph method. The mathematical substantiation is based on a systematic approach to complex mathematical and logical modeling in the wind and solar electric systems as dynamic systems. Decision-making to support the operation of energy systems is based on a mathematical description of the architecture of wind and solar electric systems, the methodology of the mathematical description of the dynamics of energy systems, and the cause-effect graph method. Coordination of energy production and consumption is based on forecasting changes within the parameters of technological processes. Real wind and solar electric systems are dynamic systems, the mathematical models of which reflect the properties of the transformation of influences, that is, their dynamic properties. The maintenance of the operation of wind and solar electric systems takes place in the composition of such wind and solar electric systems, which are based on dynamic systems. When designing wind and solar systems, we lay integrated dynamic subsystems for evaluating changes in both energy production and consumption at their foundations. By representing the design of technological systems as an organization of complex systems, we expand it by building dynamic subsystems on its foundation – nodes that predict the components of the technological processes. Based on the structured system and mathematical substantiation of the architecture of the wind and solar electric systems, we consider the relation category as organizing interactions not only within the elements of the technological systems but also within the elements of the integrated dynamic subsystems, which makes it possible to perform operability control and identify the state of the energy systems based on the developed method of the cause-effect graph. The final assessment, obtained by using logical relations in the dynamic subsystems, enables the use of new properties of the energy systems as a result of appropriate decision-making and identification of new operating conditions. Moreover, the relations between the dynamic subsystems and the blocks in their composition enable, based on the assessment of the state of the parameters diagnosed in these blocks, to confirm the new conditions of the energy systems functioning.

Keywords: Smart Grid technologies, forecasting changes in process parameters, harmonization of energy production and consumption

Introduction

Distributed generation of electrical energy using renewable sources is needed due to daily and seasonal fluctuations in energy production, as well as uneven consumption. The process of transition to alternative energy – biofuel, solar, and wind, due to the stochastic nature of energy production, requires intelligent systems for managing the flow of electric energy and consumption such as Smart Grid technologies. Demand management systems and energy storage are new components for the integration of distributed energy generation in the energy system. An urgent, further development in this direction is predicting changes in technological process parameters so as to coordinate production and energy consumption. The materials presented are the result of research carried out based on a systematic approach. Mathematical description of the architecture of electric systems, methodology of mathematical description of power systems dynamics, cause-effect graph method, and mathematical description of Smart Grid are offered and tested the supporting of electric systems operation. Integrated Smart Grid systems are developed for the coordination of the production and consumption of electric power based on the forecasting of changes in parameters of technological processes.

Published Monograph works: 1. E. Chaikovskaya (2022). Book. Smart Grid Technologies in Electric Systems for Renewable Energy. Energy Science, Engineering and Technology. ISBN 979-8-88697-387-7 Published by NOVA Science Publishers Inc. New York. P. 181. (Scopus)1. 2. E. Chaikovskaya (2021). Book. Chapter 6. Smart Grid Technology for Maintaining the Functioning of a Wind-Solar Electric System. Advances in

Energy Research. Volume 35. Morena J. Acosta (Editor). NOVA Science Publishers Inc. New York. P. 215-236.

The presented materials have been reviewed at international scientific conferences and in scientific journals, and referenced in the Scopus database (10 personal publications) while the author was teaching the following subjects at the Odessa Polytechnic National University: "A systematic approach in the study of non-traditional energy objects"; "Wind energy installations"; "Humidification and diagnostics of energy equipment"; "Mathematical problems of renewable energy"; "Production and use of biofuel"; "Energy technological installations"; "Systems and modes of energy supply"; "Computer technology and algorithmic languages." The author supervised the preparation of bachelor's and master's theses, as well as three dissertations of post-graduate students. One of the dissertations was defended in front of the Dissertation Defense Council. The applicant was awarded the degree of candidate of technical sciences. The author published 247 printed scientific works in national as well as international databases. The author is cited in scientific articles in foreign scientific journals. The author is rated for the work carried out.

Methodological and Mathematical Substantiation of the Architecture of the Wind and Solar Electric Systems

One of the main properties of energy systems is the mandatory exchange of substances, energy, and information with the environment. The functioning of energy systems can be considered, in this regard, as a reproduction of external and internal influences and changes in initial conditions. The nature of the reaction is determined by the inertia of the devices and the speed of transient processes, that is, by the dynamic properties that appear during the operation of energy systems, information about which can be obtained from the dynamic characteristics. The dynamic characteristics of energy systems can be described by a finite set of parameters regarding changes in time and a spatial coordinate that coincides with the direction of the medium's flow. Input influences can also be described by a set of parameters. Therefore, the dynamic description of the energy system most fully and multifacetedly characterizes its operation. Thus, we determine that a real energy system is a dynamic system, the mathematical model of which reflects the properties of the transformation of influences that is its dynamic properties. The operational maintenance of energy systems takes place in the composition of such technological systems, the basis of which is the dynamic system. When designing a technological system, we lay an integrated dynamic subsystem for evaluating changes in both production and energy consumption at its foundation (Figure 1.1).

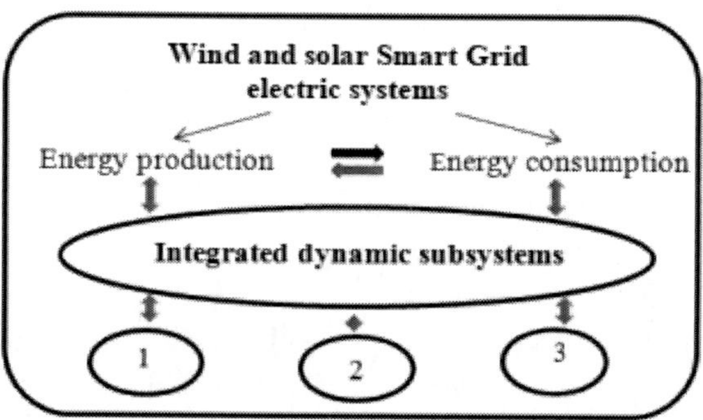

Figure 1.1. Architecture of the wind and solar electric systems: 1 – units for technological process support; 2 – units for changing operational conditions; 3 – units for estimation of functional efficiency.

Representing the construction of a technological system as an organization of a complex system, we expand it by building an integrated dynamic subsystem around its base, as well as other modules predicting the basic components of the technological process. (Figure 1.1). The relationship of the dynamic subsystem with other modules of the technological system, which assess the support of the technological process, prevent its violation, and maintain functional efficiency in decision-making conditions, is based on mathematical modeling of their logical connections, which change over time. This provides an opportunity to set new properties of the integrated dynamic subsystem – the energy system and modules of the technological system. Mathematical understanding of the principle of system organization is based on the application of logical structures that describe "organizing" the relationships between system elements. This provides an opportunity to evaluate the organization's processes using structured systems analysis. On the basis of this provision, we consider this category of relations to be cause-and-effect relationships that lead to the coordinated behavior of the dynamic subsystem and other elements of the technological system.

The methodological justification of the proposed architecture of the technological system can be presented as follows:

{(technological system), (integrated dynamic subsystem) → dynamic properties → logical relations → properties (state identification) → (integrated dynamic subsystem) → logical relations → properties (modules of technological systems) → (technological system)}.

The mathematical substantiation of the architecture of the wind and solar electric system (1) is presented here: (Figure 1.2).

Wind and solar Smart Grid electric systems

$$WSESSG(\tau) = \left\{ \begin{array}{l} \left[\begin{array}{l} ID(\tau)(P(\tau)\left\langle \begin{array}{l} x_0(\tau), x_1(\tau), x_2(\tau), f(\tau), K(\tau), \\ y(\tau, z), d(\tau), FI(\tau) \end{array} \right\rangle, LC(\tau), \\ \langle f(\tau), K(\tau), y(\tau, z), d(\tau), FI(\tau) \rangle LS(\tau), P(\tau)), \end{array} \right] \\ R(\tau), (P_i(\tau)\langle x_1(\tau), f_i(\tau), K_i(\tau), y_i(\tau), FI(\tau) \rangle LS(\tau)) \end{array} \right\} \quad (1)$$

Figure 1.2. Mathematical substantiation of the architecture of the wind and solar electric systems.

Where $WSESSG(\tau)$ – wind and solar electric systems; τ – time, seconds; $ID(\tau)$ – integrated dynamic subsystems; $P(\tau)$ – properties of the components of the wind and solar electric systems; $x(\tau)$ – impacts; $f(\tau)$ – parameters that are measured; $K(\tau)$ – coefficients of the mathematical description of the dynamics of changes in the predicted parameters; $y(\tau, z)$ – predicted output parameters; z – length coordinate coinciding with the direction of the flow of the medium, meters; $d(\tau)$ – dynamic parameters of predicted parameter changes; $FI(\tau)$ – functional resulting information; $LC(\tau)$ – logical relations regarding the control of the operational efficiency of the wind and solar electric systems; $LS(\tau)$ – logical relations regarding the identification of the state of the wind and solar electric systems; $R(\tau)$ – logical relations in $WSESSG(\tau)$ to confirm the correctness of the decisions taken by the blocks in the $WSESSG(\tau)$. Indices: i – the number of wind and solar electric systems elements; 0, 1, 2 – initial stationary mode, external and internal nature of influences.

The Methodology of Mathematical Description of the Dynamics of the Wind and Solar Electric Systems

Forecasting the parameters of the technological process to support the functioning of the technological system at the decision-making level requires an important condition. It is the movement of the system taking place in the non-linear region of its space. Irreversible thermodynamic processes taking place in a dynamic energy system should be represented by nonequilibrium thermodynamic equations, where state parameters are considered to be continuous functions of spatial coordinates and time. The nonlinearity of the dynamic processes occurring in the technological system determines the mathematical modeling of the dynamics relative to the significant parameters that are predicted. These are parameters selected based on expert knowledge, the properties of which are decisive for the functioning of this energy system. The system of differential equations includes the equation of state as an estimate of the physical model of the system, the energy equations of the transmitting and receiving media, and the heat balance equation for the heat exchanger walls. If necessary, the system of differential equations can be supplemented with a continuity equation, if the change in the flow rate of the working environment is chosen as an essential parameter to be predicted. The development of a mathematical model begins with an assessment of the physical model of the system expressed by the equation of state. Furthermore, in developing a mathematical model, the equations of state are supplemented with energy equations of the transmitting and receiving media, which are considered as equations of enthalpy transfer. Equations of the energy of the receiving environment, characterized by the presence of a significant predicted parameter, must be drawn up with a representation of the change of these parameters not only in time but also along the spatial coordinate that coincides with the direction of the medium flow.

A distinctive feature of mathematical models lies precisely in these energy equations since they contain terms that reflect a wide range of influences coming to the input of a functioning energy system from the environment. There is a need to obtain an estimate of the change of a significant parameter, which is predicted based on a wide range of disturbances coming from the environment. In this case, the result of the solution of the system of differential equations is the transfer functions obtained based on the Laplace transform, which evaluates the change of the essential parameters that are predicted when the parameters of the technological process change. Analyzing the obtained mathematical model, we set the internal parameters that are part of the coefficients of the dynamic equations. The linearization performed for the solution of the system of nonlinear differential equations is correct. In the real conditions of the energy system's operation, during the transition from stationary states and the arrival of external and internal influences, the coefficients of the dynamics equations are rearranged in time due to the change of the internal parameters that are measured. Thus, according to the mathematical justification of the architecture of technological systems, the change in the properties of the dynamic subsystem is due to the change in the initial conditions of operation. This is the reason for obtaining an estimate of the state of the internal parameter to be measured, the coefficient of the transfer function, the significant parameter to be predicted, and the dynamic parameters of the dynamic characteristics of the significant parameter to be predicted. The final assessment obtained by using logical relations in the dynamic subsystem enables the obtainment of new properties of the energy system as a result of appropriate decision-making and identification of new operating conditions. Moreover, the relations between the dynamic subsystem and the blocks in its composition allow, based on the assessment of the state of the parameters diagnosed in these blocks, to confirm the new conditions of the energy system's operation.

Cause-and-Effect Graph Method

Based on the structured system and mathematical substantiation of the architecture of the wind and solar electric system (1), we consider the relation category to be the organizing interactions not only within the elements of the wind and solar electric system but also within the elements of the integrated dynamic subsystems, which makes it possible to perform operability control and identify the state of the energy system based on the developed cause-effect graph method (Figure 1.3).

Thus, block CT_1 evaluates the change in the initial conditions of functioning due to the appearance of influences. Next, using a chain of cause-and-effect relationships, when the previous evaluation of an event that occurs is the reason for the acquisition of the next one, we receive summary information from the CTc control unit. Adopting anticipatory decisions provides an opportunity to maintain the technological process or to change the operating conditions to prevent disruption of the technological process. Moreover, there is an opportunity to functionally evaluate the change in the efficiency of the technological process and the efficiency of decision-making. After the appropriate decisions have been made, the new energy system operating conditions are confirmed using the second part of the graph of cause-and-effect relationships concerning the predicted parameters, estimated in accordance with the first part of the graph. The anticipatory decisions go through decision-making conf-irmation, which takes place based on logical connections in the dynamic subsystems, and are finalized if the dynamic subsystems receive confirmatory evaluations of the correctness of decision-making from the corresponding blocks that are a part of the wind and solar electric systems.

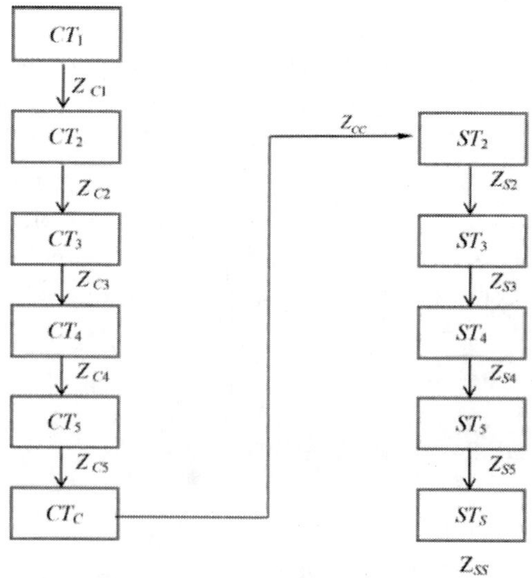

Figure 1.3. Graph of cause-and-effect relationships of the integrated dynamic subsystems: *CT*–event control; *Z*–logical relations; *ST* – event identification. Indexes: 1 – influences; 2 – internal parameters that are measured; 3 – coefficients of dynamics equations; 4 – essential parameters that are predicted; 5 – dynamic parameters; *c* – operability control; *s* – state.

Mathematical and Logical Substantiation of the Smart Grid Technologies to Maintain the Operation of the Wind and Solar Electric Systems

The proposed mathematical substantiation of operational maintenance of the wind and solar Smart Grid electric systems (2), (Figure 1.4) is based on the mathematical description of the architecture of the wind and solar electric systems (1) (Figure 1.2.), methodology of the mathematical description of dynamics of power systems, and the cause-effect graph method (Figure 1.3).

Wind and solar Smart Grid electric systems

$$WSESSG(\tau)=\begin{bmatrix}\begin{bmatrix}(ID(\tau)(P(\tau),\\ CMM(\tau,z)(sd(\tau),lp(\tau),lf(\tau),fd(\tau),tf(z,\tau),AI(\tau,z)),\\ C(\tau),LC(\tau)\\ \langle x_0(\tau),x_1(\tau),x_2(\tau),f(\tau),K(\tau),y(\tau,z),d(\tau),FI(\tau)\rangle,\\ LMD(\tau),MD(\tau),NC(\tau),S(\tau),LS(\tau)\\ \langle f(\tau),K(\tau),y(\tau,z),d(\tau),FI(\tau)\rangle P(\tau))),\end{bmatrix}\\ R(\tau),(P_i(\tau)\langle x_1(\tau),f_i(\tau),K_i(\tau),y_i(\tau),FI(\tau)\rangle LS(\tau))\end{bmatrix} \quad (2)$$

Figure 1.4. Mathematical substantiation for supporting the functioning of the wind and solar electric systems.

Where $WSESSG(\tau)$ – operational maintenance of the wind and solar Smart Grid electric system; τ – time, s; $ID(\tau)$ – integrated dynamic subsystems; $P(\tau)$ – the properties of the elements of the integrated dynamic subsystems, units of the wind and solar electric systems; $CMM(\tau, z)$ – complex mathematical modeling of the dynamics of changes predicted parameters; $sd(\tau)$ – the input data; $lp(\tau)$ – the boundary change in parameters; $lf(\tau)$ – the levels of operation; fd – the obtained parameters; $tf(\tau,z)$ – the transfer function of predicted parameters; $AI(\tau,z)$ – the standard information regarding the evaluation of the maximum admissible change predicted parameters; $C(\tau)$ – the control of operability of the wind and solar electric systems; $LC(\tau)$ – the logical relations of the control of the wind and solar electric systems; $x(\tau)$ – impacts; $f(\tau)$ – the measured parameters; $K(\tau)$ – the coefficients of the mathematical description of the dynamics of changes predicted parameters; $y(\tau, z)$ – the output predicted parameters; z – length coordinate coinciding with the direction of the flow of the medium, meters; $d(\tau)$ – the dynamic parameters of estimated change in the predicted parameters; $FI(\tau)$ – functional resulting information on decision-making; $LMD(\tau)$ – the logical relations of decision-making; $MD(\tau)$ – decision-making; $NC(\tau)$ – the new conditions of the wind and solar electric systems; $S(\tau)$ – the identification of the state of the wind and solar electric systems;

LS(τ) – the logical relations of identification of the state of the wind and solar electric systems; *R*(τ) – the logical relations between the dynamic subsystems and units as part of the wind and solar electric systems. Indices: *i* – the number of elements of *WSESSG*(τ); 0, 1, 2 – the initial, external, and internal character of influences.

Mathematical substantiation of the architecture of the wind and solar electric systems (1) (Figures 1.1, 1.2) and mathematical substantiation of maintenance of the operation of the wind and solar Smart Grid electric systems (2) (Figures 1.3, 1.4) make it possible to maintain the operation of the wind and solar electric systems using the following actions:

- Operability control (*C*(τ)) of the dynamic subsystems based on complex mathematical (*CMM*(τ, *z*)) and logical (*LC*(τ)) modeling regarding obtaining a standard (*AI*(τ,*z*)) estimate of a change in the predicted parameters;
- Operability control (*C*(τ)) of the dynamic subsystems based on complex mathematical (*CMM*(τ, *z*)) and logical (*LC*(τ)) modeling regarding the obtaining functional (*FI* (τ)) estimate of a change in the predicted parameters;
- Making anticipatory decisions to support the operation of the wind and solar electric systems (*MD*(τ)) with the use of the resulting operational information (*FI* (τ)), obtained based on logical modeling (*LMD*(τ));
- Identification (*S*(τ)) of the new conditions of the wind and solar electric systems operation (*NC*(τ)) based on logical modeling (*LS*(τ)) as a part of the dynamic subsystems and confirmation of new operating conditions based on logical modeling (*R*(τ)) from the units of the wind and solar electric systems.

Obtaining the resulting information based on energy system performance control enables the making of anticipatory decisions on the establishment of new operating conditions to be confirmed based on identification of information in the dynamic subsystems and blocks in the wind and solar electric systems.

Results and Discussion

So, for example, using methodological and mathematical substantiation (1), (2), (Figures 1.1 – 1.4) the author, Chaikovskaya, E. (2017) proposes an energy-saving technology to support changes in the battery capacity. The basis for coordinating the processes of heat and mass transfer during the charging and discharging of the battery is predicting the change in the charge and discharge voltage when measuring the temperature of the electrolyte at the inlet and outlet of the battery. For this purpose, a mathematical model of battery voltage change dynamics during the charging and discharging has been developed. The transfer functions for the channels "charge voltage–electrolyte temperature within the battery volume" and "discharge voltage–electrolyte temperature within the battery volume" make it possible to predict the change in the charge and discharge voltage based on an assessment of the change in the electrolyte temperature in the pores of the plates and above the plates. Transfer functions are obtained by solving a system of nonlinear differential equations using the Laplace transform method. The system of differential equations includes the equation of state as an estimate of the physical model of the battery, the equations of charge and discharge energy, as well as the heat balance equation for battery plate walls. The charge and discharge energy equations were developed using an estimate of the change in the electrolyte temperature in the pores of the plates and above the plates both in time and along the spatial coordinate of the battery plates. The integrated voltage change system, obtained on the basis of the coordination of heat and mass transfer processes during the charging and discharging makes it possible to make timely decisions on recharging the battery so as to prevent overcharging or unacceptable discharge. The electrochemical and diffusion processes, which accompany charging and discharging the battery, were aligned to make preemptive decisions to support a change in the capacity of the accumulator battery and to prevent the formation of gas. An analytical estimation of the change in voltage and discharge voltage without connecting the load and when connecting the load was obtained. The functional system of change in the voltage of the accumulator battery was developed, which makes it possible to maintain the capacity of the battery based on the prediction of change in the discharge voltage when measuring the temperature of electrolyte within the volume of the battery by switching to charge. This makes it possible to maintain the capacity and prevent overcharging and unacceptable battery discharge. Determining the exact time of charge before the start of gas formation makes it possible to reduce the charge time to save electricity and prevent the formation of gas. In the operation of a wind power plant with power of, for example, 10 kW, shortening the charge period and preventing gas formation reduces the cost of energy production and the payback period of the wind power plant by up to 25%. Thus, for example, using methodological and mathematical substantiation (1), (2), (Figures 1.1−1.4) the author, Chaikovskaya, E. (2019) proposes an integrated system to maintain the operation of a wind-solar electric system, based on adjusting the generation and consumption of energy. Typically, a hybrid charge controller included in a wind-solar electric system maintains the charge of a rechargeable battery by using a thermoelectric battery as a non-regulated ballast. Resetting excess energy to the ballast when using the MPPT function of the controller leads to unrecoverable losses of electrical energy, which does not make it possible to ensure an appropriate level of capacity for the charge of a rechargeable battery. Moreover, the use of a thermoelectric battery as a ballast eliminates the need to maintain the operation of a wind energy installation by using the accumulation of heat to regulate the power of a wind turbine. Failure to account for this property of a thermoelectric battery could lead to the uncontrolled acceleration of the wind turbine at high wind speeds and its malfunction. It is known that a thermoelectric battery is controlled in line with the thermostat principle, that is, when establishing the required temperature of heated local water, the thermoelectric battery is disconnected from power. Not using a change in the local water flow rate during the period of thermoelectric battery charging, when changing capacity, prolongs the charging duration and leads to considerable expenditure of electricity. A technique to overcome these difficulties has been proposed. It is the thermoelectric battery that must become the main center of adjusting a change in the total power of a wind-solar electric system to power consumption by redistributing the accumulated heat and electric energy in terms of consumption. It has been proposed to predict a change in the capacity of a rechargeable battery when measuring the total voltage at the input to a hybrid charge controller and voltage at the output from the inverter to estimate the ratio of electrical energy generation and consumption when measuring the voltage frequency. For this purpose, as a result of solving a system of non-linear differential equations, transfer functions were obtained for the channels "accumulator battery capacity – power of the thermal accumulator," "circulation pump electric motor speed – voltage frequency," "local water consumption – circulation pump electric motor speed." The system of differential equations includes the equation of state as an estimate of the physical model of the electrical system, the energy equation of the transmitting and receiving media – the heater of the thermal accumulator and local water, respectively, and the heat balance equation for the thermal electric heater walls. So as to estimate the change in the flow of local water, the system of differential equations is supplemented with a continuity equation. The equation for the energy of the receptive medium was developed based on an estimate of the change in the temperature of the local water both in time and along the spatial coordinate coinciding with the direction of the flow of the medium. Making preliminary decisions on changing the power of a thermoelectric battery makes it possible, while maintaining the capacity of the rechargeable battery, to ensure a change in the temperature of the water, based on changing the number of rotations of the circulating pump's motor, thereby reducing the charge duration by up to 30%.

Chapter 2

Smart Grid Technology for Maintaining the Functioning of Accumulator Battery

Abstract

On the basis of mathematical and logical modeling, a technological support system was developed for changing the battery capacity created based on the predicted voltage variation by measuring the temperature of electrolyte within the battery volume as a part of a technological system for battery operation. The developed technology makes it possible: to control the operational capacity of the accumulator battery in order to obtain a functional assessment of change in the total charge and discharge voltage; to obtain an integrated reference estimation of change in the charge and discharge voltage; to develop an integrated system for assessing any change in the voltage of the battery, which enables maintaining the capacity of the accumulator battery by measuring the temperature of electrolyte at the input to the battery. The maximum electrolyte temperature change – 35°C, was set when charging from a direct current supply, and a limit in voltage change for a further charge and discharge was established with a change in the consumption of electric energy. The use of an integrated system for the estimation of voltage change obtained on the basis of alignment between electrochemical and diffusion processes of discharge and charge enables the making of timely decisions on recharging so as to prevent overcharging and unacceptable discharge. Coordination of the electrochemical and diffusion processes that accompany the charging and discharging of the battery makes it possible, for example when operating a wind power plant with a capacity of 10 kW, to reduce the cost of energy production and the payback period of the wind power plant by up to 25% due to a reduction of the charge period and preventing the formation of gas.

Keywords: rechargeable battery, forecasting changes in voltage variation, harmonization of electrochemical and diffusion processes

Introduction

The use of alternative energy sources with the aim of natural fuel saving and reduction of harmful emissions into the atmosphere requires the coordination of production of energy and consumption based on accumulation (the author's works, 2016). For example, in the paper (P. Palacky, K. Baresova, M. Sobek, Ales Havel, 2016) it was proposed to manage the accumulation of electric energy by using control over energy flows with regard to their redistribution using a specially designed measuring device, which uses semiconductor transducers. The known operating modes of accumulator batteries are the following: buffer, cyclic, and mixed. Buffer mode, in which the accumulator battery is constantly charged, that is, it operates in the mode of constant recharging, leveling off uneven electricity consumption. The cyclic mode of operation that supports the discharge-charge cycle enables connecting the load after the battery charging is over. There are various means of charging lead-acid batteries: charge by stabilized current, charge by stabilized voltage, two-stage charge by stabilized current and voltage, accelerated charge, charge by asymmetric current, etc. When charging a battery with a constant charging current, sulfuric acid is additionally formed due to the transition of lead sulfate to lead dioxide and spongy lead, which is accompanied by an increase in the density of electrolyte in the pores of the plates and above the plates. Using a constant charge current, after reaching a certain charge voltage, the value of which is difficult to determine, the decomposition of water into oxygen and hydrogen takes place, which characterizes the boiling of the electrolyte. Long-term gas emissions not only cause unnecessary energy expenditure but can also damage the battery plates. During discharge, the active mass of both electrodes is converted into lead sulfate. This reaction consumes sulfuric acid and forms water. Due to this, the density of the electrolyte in the pores of the plates and above the plates decreases as the discharge progresses. Measuring the temperature of the electrolyte in the volume of used batteries enables the adjusting of the charge and discharge voltage using voltage measurements. In order to prevent overcharging and unacceptable battery discharge, support for the electrochemical and diffusion processes that accompany the charging and discharging of the battery should happen in coordinated cooperation. Forecasting a change in the voltage of charge and discharge based on the estimation of change in the temperature of electrolyte in the pores of plates and over plates while measuring the electrolyte temperature within the volume of batteries makes it possible to change the battery's operating modes in a timely manner. This substantiates the relevance of the present work. The means of optimizing the process of accumulator battery charge is based on the improvement of intelligent charge control systems with the use of special charge controllers. In the paper (P. Dost, M. Martin, C. Sourkounis, 2014) it was concluded, on the basis of the developed model of the electric accumulator battery, that it is necessary to create a specialized control system for each type of battery in connection with the effect of the battery design on the charge parameters. Thus, in the article (W. H. Chen, Y. B. Che, 2014), a special system to control accumulator batteries based on the real-time monitoring of voltage and current was presented. The control system prevents the overcharging of each battery element, but it is based on voltage measurements. For example, the paper (W. S. Wen, L. Wang, 2013) provides an intelligent battery charge support system based on the determination of the limit charge current using a segmentation method, which is devoted to enhancing the accumulator battery charge control system, but the boundary of the change of charge current in the battery charge process was not determined. In the paper (Y. L. Gong, H. Z. Li, M. Q. Li, W. D. Zhan, 2014), an intelligent system of battery charge was presented, based on information using voltage sensors, battery current, and electrolyte temperature, but it focused on the measurements of charge parameters and utilized electrolyte temperature measurement only for the purpose of temperature compensation of voltage change. In order to optimize the maintenance of the accumulator battery recharging under conditions of a partial discharge, the article (X. Yin, F. Zhang, X. Wang, Y. Jiao, Z. Ju, 2015) presented experimental methods to determine the capacity of the accumulator battery remaining after discharge, but the methods are based on real-time measurements. In the paper (R. Swathika, R. K. G. Ram, V. Kalaichelvi, R. Karthikeyan, 2013) it was concluded that the traditional control systems for a change in the charge parameters based on voltage measurements and associated with the change of external factors, such as the change of solar radiation when using photovoltaic systems, are not effective. In connection with the influence of diffusion processes on the charge of a battery, a proposal was put forward to use a fuzzy logic controller, but without coordination with the electrochemical charge process (R. Tenno, E. Nefedov, 2014). Well-known intelligent charging controllers are PWM controllers and MPPT controllers. PWM controllers operate by the method of pulse-width modulation of the charge current. MPPT controllers enable tracking the point of the maximum power of the power source for matching the change in voltage with the change in current. The use of these controllers is based on parameter measurements, which makes it difficult to align the electrochemical and diffusion processes that accompany the charging and discharging of the accumulator battery. When measuring the temperature of electrolyte within the battery volume, it is necessary to determine the limit change in the temperature of electrolyte during charging with a direct current supply and to set the limit voltage change with respect to subsequent charging and discharging when there is a change in consumption of electric energy. When measuring the temperature of electrolyte within the battery volume, it is necessary to align the electrochemical and diffusion processes that accompany the charging and discharging of the battery. This can be done based on the prediction of voltage change using a change in the temperature of electrolyte in the pores of the plates and above the plates. This approach will make it possible to take preliminary steps to support a change in the capacity of the accumulator battery.

Methodological and Mathematical Substantiation

One of the main properties of energy systems is the mandatory exchange of substances, energy, and information with the environment. The operation of the accumulator battery can be considered a reproduction of external and internal influences and changes in the initial conditions. For example, changes in the production and consumption of electrical energy. The nature of the reaction is determined by the inertia of the devices and the rate of transition processes, that is, the dynamic properties. The dynamic properties of the accumulator battery are determined from the dynamic characteristics, that is, by the assessment of change charge voltage and discharge voltage depending on time. Thus, the accumulator battery is a dynamic system. Therefore, support for the accumulator battery's operation should be a part of such a technological system that is based on the dynamic system. By representing the design of the technological system as an organization of a complex system, we expand it, by building on its foundation – the dynamic subsystem – units that predict the components of the technological process: charge, recharge, discharge, and evaluation of functional efficiency. Proceeding from the structured system, substantiation of the architecture of technological systems (Chapter 1) enables the establishment of relationships of the dynamic subsystem with other units of the technological system on the basis of mathematical modeling of their logical relations which change over time. Such an approach makes it possible to set new properties of the dynamic subsystem and units of the technological system. Following the structured system substantiation of the architecture of technological systems (Chapter 1), the category of relations is considered as the organizing of interactions inside the elements of the dynamic subsystem and units of the technological system. This makes it possible to perform control over workability, conduct identification of the state of the dynamic subsystem, and confirm new operating conditions from the units of the system on the basis of the developed cause-effect graph method (Chapter 1). Based on the methodological, mathematical, and logical substantiation of the technological systems (Chapter 1) the architecture, mathematical substantiation of the architecture (1), the following mathematical substantiation of maintenance of the operation (2) of the Smart Grid accumulator battery system is proposed (Figure 2.1).

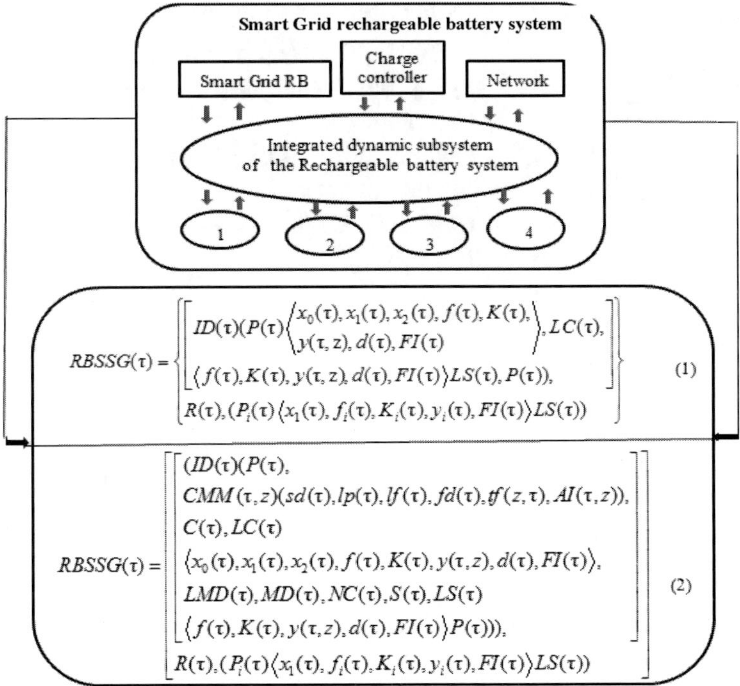

Figure 2.1. Smart Grid rechargeable battery system**:** the architecture: RB – rechargeable battery; 1 – the charging unit; 2 – the discharging unit; 3 – the recharging unit; 4 – the unit for assessing the functional efficiency. Mathematical substantiation of the architecture (1). Mathematical substantiation of operational maintenance (2).

The proposed mathematical substantiation of the architecture (1) (Figure 2.1) is based on the methodology of the mathematical description of dynamics of power systems, the cause-effect graph method (Chapter 1).

Where $RBSSG(\tau)$ – Smart Grid rechargeable battery system; τ – time, seconds; $ID(\tau)$ – integrated dynamic subsystem (charge controller and a rechargeable battery); $P(\tau)$ – properties of the components of the rechargeable battery system; $x(\tau)$ – impacts (changes in production and consumption of electric energy, etc.; $f(\tau)$ – parameters that are measured: (temperature of electrolyte in the volume of accumulator battery); $K(\tau)$ – coefficients of mathematical description of dynamics of change in the charging and discharging voltage; $y(\tau, z)$ – predicted output parameters (voltage of charge and the voltage of discharge); z – coordinate of the length of the battery plates, meters; $d(\tau)$ – dynamic parameters (voltage of charge and the voltage of discharge); $FI(\tau)$ – functional resulting information on decision-making; $LC(\tau)$ – logical relations regarding the control of the rechargeable battery system workability; $LS(\tau)$ – logical relations regarding the identification of the state of the rechargeable battery system; $R(\tau)$ – logical relations in $RBSSG(\tau)$ to confirm the correctness of decisions made from the units of the rechargeable battery system. Indices: i – the number of elements of the rechargeable battery system; 0,1,2 – initial stationary mode, external, internal nature of influences.

The proposed mathematical substantiation of operational maintenance of the Smart Grid rechargeable battery system (2), (Figure 2.1) is based on the methodology of the mathematical description of dynamics of power systems, the cause-effect graph method (Chapter 1). The proposed rationale is founded on the mathematical description of the battery system architecture (1), (Figure 2.1).

Where $RBSSG(\tau)$ – Smart Grid maintenance of the rechargeable battery system's operation; τ – time, seconds; $ID(\tau)$ – integrated dynamic subsystem (charge controller and a rechargeable battery). $P(\tau)$ – the properties of the elements of the integrated dynamic subsystem, units of the rechargeable battery system; $CMM(\tau,z)$ – complex mathematical modeling of the dynamics of changes in the charge and discharge voltage of the rechargeable battery system; $sd(\tau)$ – the input data (the rechargeable battery and its type and capacity; $lp(\tau)$ – the boundary change in parameters (the temperature of electrolyte within the accumulator battery volume); $lf(\tau)$ – the levels of the rechargeable battery operation; $fd(\tau)$ – the obtained parameters (mode parameters of the rechargeable battery); $tf(\tau,z)$ – the transfer function of predicted parameters – in the charge and discharge voltage of the rechargeable battery; $AI(\tau,z)$ – the standard information regarding the evaluation of the maximum permissible change in the charge and discharge voltage of the rechargeable battery; $C(\tau)$ – rechargeable battery system operability control; $LC(\tau)$ – the logical relations of the rechargeable battery system operability control; $x(\tau)$ – impacts (changes in production and consumption of electric energy, etc.; $f(\tau)$ – the measured parameters: (temperature of electrolyte within the accumulator battery volume); $K(\tau)$ – the coefficients of the mathematical description of the dynamics of a change in the charge and discharge voltage; $y(\tau,z)$ – the output parameters (the charge voltage, discharge voltage); z – the length coordinate of the accumulator battery plates, meters; $d(\tau)$ – the dynamic parameters of estimated change in the charge and discharge voltage; $FI(\tau)$ – functional resulting information on decision-making; $LMD(\tau)$ – the logical relations of decision-making; $MD(\tau)$ – decision-making; $NC(\tau)$ – the new conditions of the rechargeable battery system's operation; $S(\tau)$ – the identification of the state of the rechargeable battery system; $LS(\tau)$ – the logical relations in identifying the state of the rechargeable battery system; $R(\tau)$ – the logical relations between the dynamic subsystem and units for charging, discharging, recharging, and functional estimation of the rechargeable battery system efficiency. Indices: i – the number of elements of $RBSSG(\tau)$; 0,1,2 – the initial, external, and internal nature of influences.

Mathematical substantiation of the architecture of the Smart Grid rechargeable battery system (1) and mathematical substantiation of operational maintenance of the Smart Grid rechargeable battery system (2) (Figure 2.1) enables maintaining the operation of the rechargeable battery system by taking the following actions:

- operability control ($C(\tau)$) of the dynamic subsystem based on complex mathematical ($CMM(\tau,z)$) and logical ($LC(\tau)$) modeling used for obtaining the standard ($AI(\tau,z)$) estimate of a change in the charge and discharge voltage of the rechargeable battery system;
- operability control ($C(\tau)$) of the dynamic system based on complex mathematical ($CMM(\tau,z)$) and logical ($LC(\tau)$) modeling used for obtaining functional ($FI\ (\tau)$) estimate of a change in the charge and discharge voltage of the rechargeable battery system;
- decision-making ($MD(\tau)$) with the use of the resulting functional information ($FI\ (\tau)$) obtained based on logical modeling ($LMD(\tau)$); to make timely decisions on recharging so as to prevent overcharging or unacceptable discharge;

- identification ($S(\tau)$) of new conditions in the operation of the rechargeable battery system ($NC(\tau)$) based on logical modeling ($LS(\tau)$) as a part of the dynamic subsystem and confirmation of new operating conditions based on logical modeling ($R(\tau)$) from the units of the rechargeable battery system.

Maintaining the Battery Capacity Based on a Prediction of Changes in Voltage

The basis for the alignment of electrochemical and diffusion of charging and discharging processes in the accumulator battery is the prediction of change in the charge and discharge voltage while measuring the temperature of electrolyte at the input and output of the accumulator battery. For this purpose, a mathematical model is constructed of the dynamics of change in the accumulator battery voltage during charging and discharging.

According to formulas (1) and (2), a transfer function of the "charge voltage – temperature of electrolyte within the accumulator battery volume" relationship, which enables predicting changes in the charging voltage based on the assessment of change in the temperature of electrolyte within the plate pores and over the plates, both over time and along the length of the battery plates, is represented in the following way:

$$W_{u-t_1} = \frac{K_{\text{t}} K_{\text{u}} \varepsilon \left(1 - L^*_{\text{ch}}\right)}{(T_{\text{ch}} + 1)\beta - 1}\left(1 - e^{-\gamma\xi}\right), \tag{3}$$

where

$$K_t = \frac{m\left(\theta_0 - t_{\text{e}2}\right)}{G_{\text{ch}0}};$$

$$K_{\text{u}} = (0.84+(\rho+0.0007(t_{\text{e}1}-t_{\text{e}25})+I_{\text{ch}} R_{\text{ch}} n;$$

$$\varepsilon = \frac{\alpha_{\text{disch}\,0} h_{\text{disch}\,0}}{\alpha_{\text{ch}\,0} h_{\text{ch}\,0}};$$

$$L_{ch}^{*} = \frac{1}{L_{ch}+1}; \quad L_{ch} = \frac{G_{ch}C_{ch}}{\alpha_{ch0}h_{ch0}}; \quad \gamma = \frac{(T_{ch}S+1)\beta - 1}{\beta};$$

$$\xi = \frac{z}{L_{h}}; \quad T_{ch} = \frac{g_{ch}C_{ch}}{\alpha_{ch0}h_{ch0}};$$

$$\beta = T_{m}S + \varepsilon^{*} + 1; \quad T_{m} = \frac{g_{m}C_{m}}{\alpha_{ch0}h_{ch0}}; \quad \varepsilon^{*} = \varepsilon(1 - L_{ch}^{*}),$$

where α is the heat emission coefficient, kW/(m^2·K); C is the specific heat capacity, kJ/(kg·K); G is the consumption of substance, kg/s; g is the specific mass of substance, kg/m; h is the specific surface, m^2/m; t, θ is the electrolyte temperature within the battery volume and the plate wall, respectively, K; t_{e1}, t_{e2} are the temperatures of electrolyte at the battery input and the battery output, respectively, K; t_{e25} is the temperature of electrolyte within the battery volume at 25°C outside temperature; U is the total voltage, W; R is the battery resistance, Ohm; z is the length coordinate of the accumulator battery plates, meters; T, T_m are the constants of time, characterizing thermal accumulation ability of the electrolyte, metal, seconds; m is the indicator of heat emission coefficient dependence on consumption; ι is time, seconds; ρ is the density of electrolyte, g/m^3; n is the number of accumulators in the battery; I is current, A; S is the Laplace transform parameter; $S = \omega j$; ω is frequency, 1/s. Indices: 0 – initial stationary mode; 1 – input to the accumulator battery; ch, disch. are the charge and discharge of the accumulator battery, respectively; m is the metal wall.

The transfer function for the "charge voltage – electrolyte temperature within the accumulator battery volume" relationship (3) is obtained by solving the system of nonlinear differential equations using the Laplace transform. The system of differential equations includes the state equation as an estimate of the physical model of the accumulator, the equation of charge and discharge energy, and the equation of heat balance for the accumulator plate wall. The charge energy equation was developed by using changes in the temperature of the electrolyte within the plate pores and above the plates not only over time but also along the spatial coordinates of accumulator plates. The charge energy equation includes K_u coefficient, which estimates a change in the accumulator charge voltage when the electrolyte temperature at the input to the accumulator battery changes.

According to formulas (1) and (2), the transfer function for the "discharge voltage – electrolyte temperature within the accumulator battery volume" relationship, which enables the predicting of a change in the discharge voltage based on the estimation of change in the temperature of electrolyte in the pores of the plates and above the plates not only over time but also along the spatial coordinates of accumulator plates, is represented in the following way:

$$W_{u-t_1} = \frac{K_t K_u \varepsilon\left(1 - L_{\text{disch}}^{*}\right)}{(T_{\text{disch}} S + 1)\beta - 1}\left(1 - e^{-\gamma\xi}\right), \tag{4}$$

where

$$K_t = \frac{m\left(\theta_0 - t_{e2}\right)}{G_{\text{disch}0}};$$

$$K_u = (0.84 + (\rho - 0.0007(t_{e1} - t_{25}) - I_{\text{disch}} R_{\text{disch}})n;$$

$$\varepsilon = \frac{\alpha_{\text{ch}0} h_{\text{ch}0}}{\alpha_{\text{disch}0} h_{\text{disch}0}};$$

$$L_{\text{disch}}^{*} = \frac{1}{L_{\text{disch}} + 1}; \quad L_{\text{disch}} = \frac{G_{\text{disch}} C_{\text{disch}}}{\alpha_{\text{disch}0} h_{\text{disch}0}}; \quad \gamma = \frac{(T_{\text{disch}} S + 1)\beta - 1}{\beta};$$

$$\xi = \frac{z}{L_{\text{disch}}}; \quad T_{disch} = \frac{g_{\text{disch}} C_{\text{disch}}}{\alpha_{\text{disch}0} h_{\text{disch}0}};$$

$$\beta = T_{\text{m}} S + \varepsilon^{*} + 1; \quad T_{\text{m}} = \frac{g_{\text{m}} C_{\text{m}}}{\alpha_{\text{disch}0} h_{\text{disch}0}}; \quad \varepsilon^{*} = \varepsilon(1 - L_{\text{disch}}^{*}).$$

The transfer function for the "discharge voltage – electrolyte temperature within the accumulator battery volume" relationship (4) was obtained by solving the system of nonlinear differential equations using the Laplace transform. The system of differential equations includes the state equation as an estimate of the physical model of the accumulator, the charge and discharge energy equation, and the equation of heat balance for the accumulator plate's wall. The discharge energy equation was developed by using a change in the temperature of electrolyte within the plate pores and above the plates not only over time but also along the spatial coordinates of accumulator plates. The discharge energy equation includes a K_u coefficient, which evaluates a change in the discharge voltage of the accumulator when the electrolyte temperature at the accumulator battery input changes.

A valid part of the transfer function (3) was singled out to estimate a change in the charge voltage when the electrolyte temperature changes within the accumulator volume:

$$O(\omega)=\frac{(L_1A_1)+(M_{1.}B_1)K_tK_u\varepsilon(1-L_{\text{ch}}^{*})}{(A_1^{2}+B_1^{2})}. \tag{5}$$

K_t coefficient includes the temperature of the dividing wall θ:

$$\theta=(\alpha_{\text{ch}}(\sigma_1+\sigma_2)/2)+A(t_1+t_2)/2)/(\alpha_{\text{ch}}+A), \tag{6}$$

where σ_1, σ_2 are the temperatures of electrolyte at the input and output to the accumulator, K, respectively;

$$A=1/(\delta_{\text{m}}/\lambda_{\text{m}}+1/\alpha_{\text{disch}}), \tag{7}$$

where δ is the thickness of the accumulator plate wall, meters; α is the heat emission coefficient, kW/(m^2·K); λ is the heat conductivity of the accumulator plate metal, kW/(m·K); t_1, t_2 are the temperatures of electrolyte within the plate pores and above the plates at the input and output to accumulator, K, respectively. Indices: ch – the accumulator battery charge; disch – the accumulator battery discharge; m – metal wall.

To use the valid part of $O(\omega)$, the following coefficients were obtained:

$$A_1=\varepsilon^{*}-T_{\text{ch}}T_{\text{m}}\omega^2;\;\; A_2=\varepsilon^{*}+1;\;\; B_1=T_{\text{ch}}\varepsilon\omega+T_{\text{ch}}\omega+T_{\text{m}}\omega; \tag{8}$$

$$B_2=T_{\text{m}}\omega;\;\; C_1=\frac{A_1A_2+B_1B_2}{A_2^{2}+B_2^{2}};\;\; D_1=\frac{A_2B_1-A_1B_2}{A_2^{2}+B_2^{2}}; \tag{9}$$

$$L_1 = 1 - e^{-\xi C_1}\cos(-\xi D_1);\ \ M_1 = -e^{-\xi C_1}\sin(-\xi D_1). \tag{10}$$

The valid part of the transfer function (4) was singled out to estimate a change in the discharge voltage when the electrolyte temperature changes within the accumulator volume:

$$O(\omega) = \frac{(L_1 A_1) + (M_{1.} B_1) K_t K_u \varepsilon (1 - L_{disch}^{*})}{(A_1^2 + B_1^2)}. \tag{11}$$

K_t coefficient includes the temperature of the dividing wall θ:

$$\theta = (\alpha_{disch}(\sigma_1 + \sigma_2)/2) + A(t_1 + t_2)/2)/(\alpha_{disch} + A), \tag{12}$$

where σ_1, σ_2 are the temperatures of electrolyte at the input and output to the accumulator, K, respectively;

$$A = 1/(\delta_m/\lambda_m + 1/\alpha_{ch}), \tag{13}$$

where δ is the thickness of the accumulator plate wall, meters; α is the heat emission coefficient, kW/(m^2·K); λ is the heat conductivity of the accumulator plate metal, kW/(m·K); t_1, t_2 are the temperatures of electrolyte within the plate pores and above the plates at the input and output to the accumulator, K, respectively. Indices: ch – accumulator battery charge; disch – battery discharge; m – metal wall.

The implementation of transfer functions (3) and (4) obtained by using the operator method for solving the system of nonlinear differential equations holds a Laplace transform parameter – $S(S = \omega j)$, where ω is the frequency, 1/sec. The valid parts (5) and (11) obtained from the mathematical processing of the transfer functions, were selected for the transition from the frequency region to the time region in order to obtain dynamic characteristics of the charge and discharge voltage. These parts are included in the integral (14), which enables obtaining estimates of changes in the charge and discharge voltage over time using the inverse Fourier transform. Using the integral of transition from the frequency region to the time region, a change in the charge and discharge voltage was determined in the following way:

$$U(\tau) = t(z,\tau)K_u(\tau) = \frac{1}{2\pi}\int_0^\infty O(\omega)\sin(\tau\omega/\omega)d\omega, \tag{14}$$

where t is the temperature of electrolyte within the plate pores and above the plates, K.

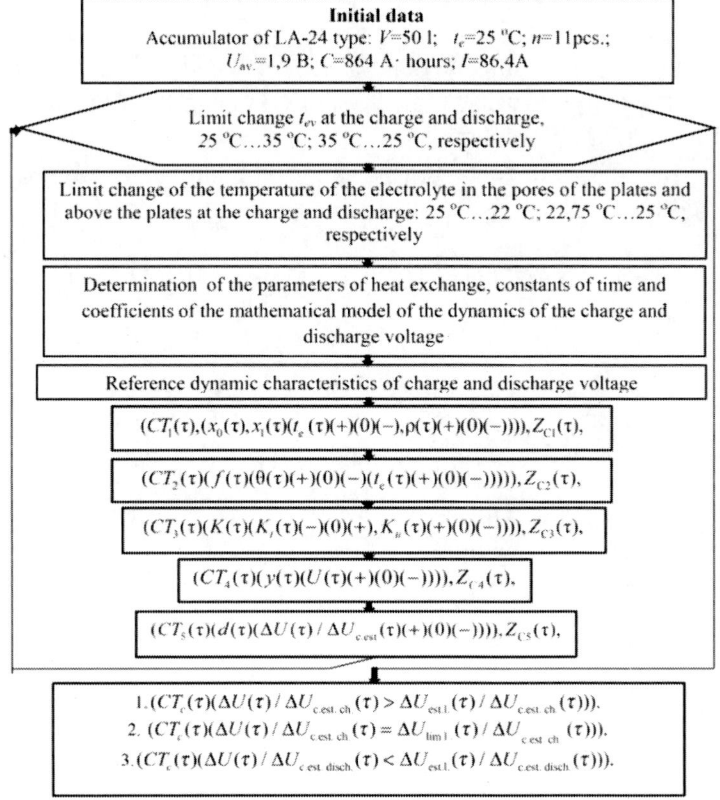

Figure 2.2. Schematic of comprehensive mathematical and logical modeling of a 12 V accumulator battery: V – volume of electrolyte – 25% solution of sulfuric acid in one accumulator, l; n – number of accumulator plates, pcs .; U – voltage, V; C – capacity, A· hours; I – current at 10-hours discharge, A; t_{ev}, t_e – temperature of the electrolyte in the volume of accumulators, at the input of the accumulator battery, respectively, K; CT – event control; Z – logical relations; d – dynamic parameters; x – impacts; f – parameters which are diagnosed; y – output parameters; K – coefficients of mathematical description; Indices: c – work-status control; mid. – the average parameter value; c.est.ch., c.est.disch. – constant estimated value of the charge and discharge parameters, respectively; est.l. – estimated value of the operational level parameter; 0,1,2 – initial stationary mode, external, internal parameters; 3 – coefficients of the dynamics equations; 4 – essential parameters that are diagnosed; 5 – dynamic parameters.

By using formulas (1–4) for comprehensive mathematical and logical modeling in order to obtain the reference and functional estimation of change in the total charge and discharge voltage, a structural circuit (Figure 2.2) was developed, according to which, at the established limit change in the temperature of electrolyte during charging and discharging, parameters of heat transfer, time constants and coefficients of the dynamics of changes during charging and discharging have been determined (Tables 2.1–2.3), the reference charge and discharge integrated system (Table 2.4, Figure 2.3) and the resulting functional information (1)–(3) on decision-making to support a change in the capacity of accumulator battery using the cause-effect graph method (Chapter 1) were obtained.

The charging voltage in the preset time period is determined in the following way:

$$U_{chi+1}(\tau) = U_{chi}+((\Delta U_{chi}(\tau)/\Delta U_{c.est.ch}(\tau) - \Delta U_{chi+1}(\tau)/\Delta U_{c.est.ch}(\tau))/(t_{ech.output}-t_{echinputi})U_{chi}(\tau))), \quad (15)$$

where index *i* is the number of charge levels.

Table 2.1. Parameters of heat exchange during charging and discharging

Level of operation	Parameter		
	α_{ch}, W/(m^2·K)	α_{disch}, W/(m^2·K)	k, W/(m^2·K)
Charge, discharge	14.179	14.185	3.25

Note: α_{ch} – coefficient of heat emission from the electrolyte to the wall of the battery plate at a charge, W/(m^2·K); α_{disch} – coefficient of heat emission from the wall of the battery plate to the electrolyte at discharge, W/(m^2·K); k – coefficient of heat transfer, W/(m^2·K).

Table 2.2. Time constants and coefficients of mathematical model of charge dynamics

Level of operation	T_{ch}, c	T_m, c	ε	ε*	ζ	L_{ch}, m	L_{ch}^*
Charge	1467.56	13352.5	1	0.973	0.647	35.51	0.027

Table 2.3. Time constants and coefficients of mathematical model of discharge dynamics

Level of operation	T_{disch}, c	T_m, c	ε	ε*	ζ	L_{disch}, m	L_{disch}^*
Discharge	1466.94	13346.86	1	0.973	0.647	35.51	0.027

Table 2.4. Reference integrated charge and discharge system of a 12 V accumulator battery

Time, τ, 10^3 sec	I_{ch} = 86.4 A $t_{e\ ch\ output}$ = 35°C Change $t_{e\ ch\ input}$, °C	$\Delta U_{ch}(\tau)$ $/\Delta U_{c.est.ch.}(\tau)$	$U_{ch}(\tau)$, V	$I_{disch.}$ = 86.4 A $t_{edischoutput}$ = 25°C Change $t_{e\ disch.\ input}$, °C	$\Delta U_{disch.}(\tau)$ $/\Delta U_{c.est.disch.}(\tau)$	$U_{disch.}(\tau)$, V
0	25	1	12.1072	35	-1	12.0486
3	25.62	0.9732	12.1396	34.375	-0.9198	11.9520
6	26.25	0.9545	12.1638	33.75	-0.8396	11.8498
9	26.875	0.9358	12.1897	33.125	-0.7593	11.7411
12	27.5	0.9170	12.2179	32.5	-0.6790	11.6251
15	28.125	0.8983	12.2485	31.875	-0.5987	11.5006
18	28.75	0.8795	12.2818	31.25	-0.5183	11.3661
21	29.37	0.8608	12.3226	30.625	-0.4379	11.2217
24	30	0.8420	12.3689	30	-0.3574	11.0611
27	30.62	0.8232	12.4154	29.375	-0.2770	10.8832
30	31.25	0.8044	12.4699	28.750	-0.1964	10.6827
33	31.87	0.7857	12.5321	28.125	-0.1158	10.4531
36	32.5	0.7669	12.6075	27.5	-0.0352	10.1835
39	33.12	0.7481	12.7023	27.2	0	10.0566
42	33.75	0.7292	12.8303	–	–	–
45	34.37	0.7104	13.0233	–	–	–
48	35	0.6916	13.4150	–	–	–

Note: $t_{e\ ch\ input}$, $t_{e\ ch\ output}$, $t_{e\ disch.\ input}$, $t_{e\ disch\ output}$ – electrolyte temperature at the input and output of the accumulator battery during charging and discharging,°C, respectively; I_{ch}, I_{disch} – charge and discharge current, A, respectively; U_{ch}, U_{disch} – the charge and discharge voltage, V, respectively. Index: c.est.ch, c.est.disch – constant estimation value of the parameter for charge and discharge, respectively.

For example, for a period of 3000 seconds (0.83 hours) from the onset of charging the accumulator battery with an electrolyte temperature at the accumulator battery input equaling 25.62°C, the absolute value of voltage during this period is determined in the following manner:

$$12.1396\ V = 12.1072\ V+(1–0.9732)/(35°C–25°C)12.1072\ V.$$

The discharge voltage in the preset time period is determined in the following way:

$$U_{disch\,i+1}(\tau) = U_{disch\,i}-((\Delta U_{disch\,i}(\tau)/\Delta U_{c.est.disch}(\tau) - \Delta U_{disch\,i+1}(\tau)/\Delta U_{c.est.disch}(\tau))/(t_{edischinput}-t_{edischioutput})U_{dischi}(\tau))), \quad (16)$$

where index i is the number of charge levels.

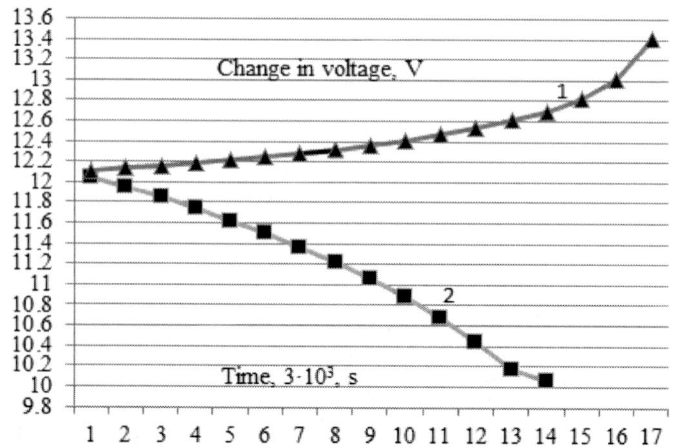

Figure 2.3. Reference integrated system of change in the voltage of a 12 V accumulator battery, where 1 – accumulator battery charge, 2 – accumulator battery discharge.

Table 2.5. A functional integrated system of change in the charge voltage of a 12 V accumulator battery within a 50% change in the capacity

Time, τ, 10^3 s	0	3	6	9	12	15	18
$t_{e\ disch\ input}$, °C	35	34.375	33.75	33.125	32.5	31.875	31.25
$\Delta U_{disch}(\tau)$ / $\Delta U_{c.est.disch}(\tau)$	–1	–0.9198	–0.8396	–0.7593	–0.6790	–0.5987	–0.5183
$U_{disch}(\tau)$, V	13.415	13.3074	13.1936	13.0725	12.9433	12.8047	12.6550

Note: $t_{e\ disch.input}$ – electrolyte temperature at the input of the accumulator battery, °C, respectively; U_{disch} – discharge voltage, V. Indices: c.est.disch – constant estimated value of the parameter at discharge.

For example, for a period of 3,000 seconds (0.83 hours) from the onset of the accumulator battery discharging with an accumulator battery electrolyte temperature equaling 34.375°C, the absolute value of voltage during this period is determined in the following way:

$$11.9520\ V = 12.0486\ V–(–0.9198)–(–1))/(35°C–25°C)12.0486\ V.$$

The accumulator battery operability control (Figure 2.2) enables obtaining the resulting information for advance decision-making about supporting a change in the capacity of the accumulator battery. Based on the proposed mathematical substantiation (1) to (4), (Figure 2.2) a structural circuit (Figure 2.4) was compiled for making decisions about supporting a change in the accumulator battery capacity using a functional integrated system for changing the accumulator battery discharge voltage (Tables 2.5, 2.6) and $\Delta U_{ch}(\tau)/\Delta U_{c.est.ch}(\tau)$ (Table 2.4).

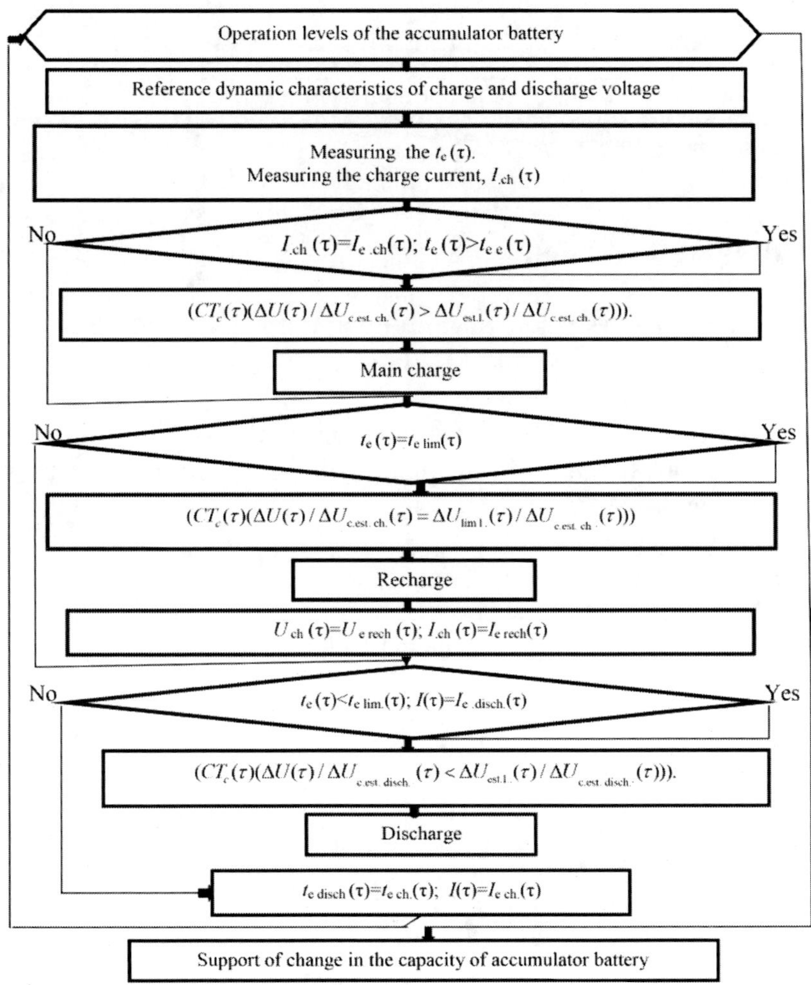

Figure 2.4. Schematic for supporting changes in the capacity of the accumulator battery, where, $I_{e\ ch}$, $I_{e\ disch}$, $I_{e\ rech}$ – the reference charging, discharging, and recharging current, respectively, A; U – voltage, V; $t_{e\ e}$, t_e – reference and functional temperature of the electrolyte within the accumulator volume, K. Indices: lim – limit change in temperature of electrolyte; ch – charge, disch – discharge; e rech – reference value of the recharge parameter; τ – time, seconds.

Table 2.6. A functional integrated system of change in the charge voltage of a 12 V accumulator battery within a 50% change in the capacity

Time, τ, 10^3 sec	21	24	27	30	33	36	39
$t_{e\ disch\ input}$, °C	30.625	30	29.375	28.750	28.125	27.5	27.2
$\Delta U_{disch}(\tau)$ $/\Delta U_{c.est.disch}(\tau)$	–0.4379	–0.3574	–0.2770	–0.1964	–0.1158	–0.0352	0
$U_{disch}(\tau)$, V	12.4925	12.3137	12.1157	11.8925	11.6399	11.3397	11.1984

Note: $t_{e\ disch.input}$ – electrolyte temperature at the input to the accumulator battery, °C, respectively; U_{disch} – discharge voltage, V. Index: c.est.disch – constant estimated value of the parameter at discharge.

Results and Discussion

The Smart Grid System of Accumulator Battery Operational Maintenance at the Decision-Making Level

An integrated system was designed to support a change in the accumulator battery capacity (Table 2.7, Figure 2.5) to predict a voltage change at continuous measurement of electrolyte temperature at the input and output of the accumulator battery.

Table 2.7. Integrated system to support the operation of a 12V accumulator battery

Time, τ, 10^3 sec	Change in the battery capacity	$\Delta U(\tau)/\Delta U_{c.est}(\tau)$	U(τ), V
0	$I(\tau) = I_{rech}(\tau)$. Recharge. $t_{e.\ input}$ = 35°C; $t_{e.output}$ = 35°C	0.6916	13.4150
3	$I(\tau) = I_{disch}(\tau)$. Discharge. $t_{e.input}$ = 35°C; $t_{e.output}$ = 27.5°C	–1	13.4150
6	$I(\tau) = I_{disch}(\tau)$. Discharge: $t_{e.input}$ = 34.375°C; $t_{e.output}$ = 27.5°C	–0.9198	13.3074
9	$I(\tau) = I_{disch}(\tau)$. Discharge. $t_{e.input}$ = 33.75°C; $t_{e.output}$ = 27.5°C	–0.8396	13.1936
12	$I(\tau) = I_{disch}(\tau)$. Discharge: $t_{e.input}$ = 33.125°C; $t_{e.output}$ = 27.5°C	–0.7593	13.0725
15	Decision-making about charge. $I(\tau) = I_{ch}(\tau)$. $t_{e.input}$ = 32.5°C; $t_{e.output}$ = 27.5°C	–0.6790	12.9433
18	$I(\tau) = I_{ch}(\tau)$. Charge. $t_{e.input}$ = 32.5°C; $t_{e.output}$ = 35°C	0.7669	12.9433
21	$I(\tau) = I_{ch}(\tau)$. Charge $t_{e.input}$ = 33.125°C; $t_{e.output}$ = 35°C	0.7481	13.0406
24	$I(\tau) = I_{ch}(\tau)$. Charge $t_{e.input}$ = 33.75°C; $t_{e.output}$ = 35°C	0.7292	13.1720
27	$I(\tau) = I_{ch}(\tau)$. Charge $t_{e.input}$ = 34.37°C; $t_{e.output}$ = 35°C	0.7104	13.3701
30	Decision-making about recharge. $I(\tau) = I_{rech}(\tau)$. Recharge $t_{e.input}$ = 35°C; $t_{e.output}$ = 35°C	0.6916	13.4150
33	$I(\tau) = I_{rech}(\tau)$. Recharge. $t_{e.input}$ = 35°C; $t_{e.output}$ = 35°C	0.6916	13.4150

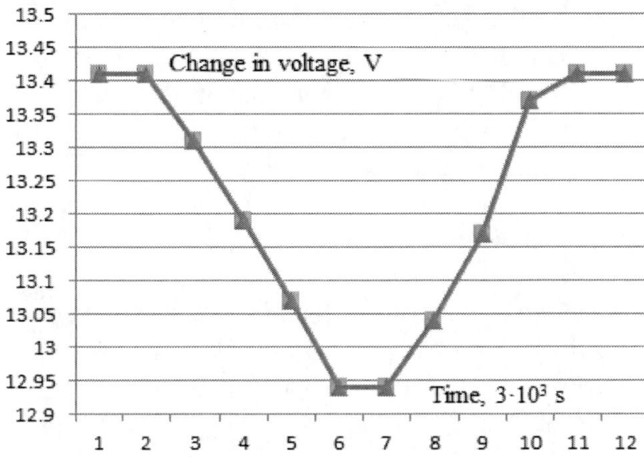

Figure 2.5. Support of change in the capacity of a 12V accumulator battery.

For example, over a period of $15 \cdot 10^3$ sec (4.17 hours) after recharging and the onset of the discharging of the accumulator battery, it was found that the electrolyte temperature within the accumulator battery volume decreased from 35°C to 32.5°C (Table 2.7), which may lead to a decrease in the battery capacity by more than 25%. In order to support a change in the battery capacity, it was decided to charge the battery using the data (Table 2.7) regarding the supply of charging current and restoring the battery capacity within 25% by connecting it to the direct current for further recharging. The absolute value of discharge voltage using formula (16) over a period of $15 \cdot 10^3$ is determined in the following way:

$$12.9433\text{V} = 13.0725\text{V} - (-0.6790 - (-0.7593))/(33.125°\text{C} - 25°\text{C})13.0725\text{V}$$

Making a decision about charging the accumulator battery over a time of $15 \cdot 10^3$ sec (4.17 hours) is confirmed by a change in the charge voltage, which has the following absolute value calculated using formula (15):

$$13.0406\text{V} = 12.9433\text{V} + (0.7669 - 0.7481)/(35°\text{C} - 32.5°\text{C})12.9433\text{V}.$$

Thus, the prediction of a change in the discharge voltage of the accumulator battery makes it possible to make preliminary decisions about charging the accumulator to support a change in capacity.

The limit change of the electrolyte temperature at the charge for direct current supply was determined and the limit voltage change for the subsequent recharging and discharging when the electric energy consumption changes was established. It is proposed to measure the electrolyte temperature within the accumulator' volume to predict a voltage change using the estimation of change in the electrolyte temperature within the plate pores and above the plates. The electrochemical and diffusion processes that accompany charging and discharging the battery were aligned so as to make preemptive decisions to support a change in the capacity of the accumulator battery and to prevent the formation of gas.

An analytical estimation of the change in voltage and discharge voltage without connecting the load and when connecting the load was obtained. The functional system of change in the accumulator battery voltage was developed, which makes it possible to maintain the battery capacity based on the prediction of change in the discharge voltage by measuring the electrolyte temperature within the accumulator volume and switching it to charge. This makes it possible to provide capacity maintenance and prevent overcharging and unacceptable battery discharge. Determining the exact duration of charging before the start of the formation of gas enables reducing the charging times to save electricity and prevent gas forming.

Chapter 3

Smart Grid Technology for Operational Maintenance of a Thermal Electric Accumulator

Abstract

By utilizing mathematical and logical modeling as part of the technological system for the operation of thermal electric accumulators, a system of technological support for changing the capacity of accumulators was developed based on predicting changes in the temperature of heated water within the accumulator volume when the temperature of water is measured at the thermoelectric accumulator's inlet and outlet. The technology developed enables controlling the operation of a thermal electric accumulator in order to obtain a functional assessment of the change in the temperature of the heated water. The use of an integrated system for estimating the change in the temperature of heated water within the battery volume, obtained by matching the heat and mass transfer processes during discharging and charging, makes it possible to make timely decisions on changing the power of a thermal electric accumulator based on a change in the local water flow rate, which enables reducing the charging time by up to 30%. For example, this technology is used in the developed integrated Smart Grid system to support the operation of a wind-solar power plant by predicting changes in the battery capacity (Chapter 6). Changes in the speed of rotation of the circulation pump electric motor are provided under the conditions of changes in the flow rate and temperature of the heated water by reducing the charging duration by up to 30%. The storage battery and the thermoelectric accumulator, as part of the network solar power system (Chapter 4), acquire the additional status of voltage regulators within the distribution system. Preliminary decisions are made to change the capacitance of a thermoelectric accumulator when redistributing the accumulated electrical energy. Changing the level of transmission of electrical energy to the grid enables maintaining the voltage within the distribution system by maintaining the power factor of the solar power plant grid. The peak load on the power system is prevented, which reduces the consumption of electricity from the network by up to 14%.

Keywords: thermal electric accumulator, Smart Grid technologies, forecasting changes in water temperature, coordination of heat and mass transfer processes

Introduction

An integrated system for operational maintenance of a wind-solar electric system has been built (Chapter 6) based on adjusting the generation and consumption of energy for the purpose of energy efficiency. Typically, a hybrid charge controller included in a wind-solar electric system maintains the charge of a rechargeable battery by using a thermoelectric battery as a non-regulated ballast. Redirecting excess energy to the ballast when using the MPPT function of the controller leads to unrecoverable losses of electrical energy, which does not make it possible to ensure an appropriate level of capacity for charging the rechargeable battery. Moreover, the use of a thermoelectric battery as a ballast eliminates the need to maintain the operation of a wind energy installation by using the accumulation of heat in order to regulate the power of a wind turbine. Failure to account for this property of a thermoelectric battery could lead to the uncontrolled acceleration of the wind turbine at high wind speeds and its malfunction. It is known that a thermoelectric battery is controlled in line with the thermostat principle, that is, when establishing the required temperature of heated local water, the thermoelectric battery is disconnected from power. Not using a change in the local water flow rate during the period of thermoelectric battery charging, when changing capacity, prolongs the duration of charging and leads to considerable expenditure of electricity. A technique to overcome these difficulties has been proposed. It is the thermoelectric battery, which must become the main center of adjusting a change in the total power of a wind-solar electric system to power consumption by redistributing the accumulated heat and electric energy in terms of consumption. It has been proposed to predict a change in the capacity of a rechargeable battery when measuring the total voltage at the input to a hybrid charge controller and voltage at the output from the inverter to estimate the ratio of electrical energy generation and consumption when measuring the voltage frequency.

Making preliminary decisions on changing the power of a thermoelectric battery makes it possible, while maintaining the capacity of the rechargeable battery, to ensure a change in the temperature of heated local water, based on changing the number of rotations of the circulating pump's electric motor in terms of water flow rate, thereby reducing the charge duration by up to 30%. The representation of the thermoelectric battery as the main center in the redistribution of the accumulated energy provides an opportunity to use the generated electrical energy completely, in terms of adjusting it to consumption.

As part of the network solar electric system (Chapter 4), the accumulator battery and the thermoelectric accumulator acquire additional status of voltage regulators in the distribution system. A comprehensive integrated system has been developed to support the functioning of the solar electric system network based on the prediction of changes in the battery capacity and power factors. Advanced decisions on changing the capacity of the thermoelectric accumulator redistribute the accumulated electrical energy. Changing the level of transmission of electric energy to the network enables maintaining voltage within the distribution system by maintaining the power factor of the solar electric system network. The voltage at the input to the hybrid inverter, the voltage at the output of the frequency converter, and within the distribution system is continuously measured. The change in the voltage ratio at the frequency converter output and within the distribution system is evaluated. The peak load on the energy system is prevented, thus reducing energy consumption from the network by 14%.

Methodological and Mathematical Substantiation

One of the main properties of energy systems is the mandatory exchange of matter, energy, and information with the environment.

In this regard, the operation of thermal electric accumulators can be considered a reproduction of external and internal influences and changes in initial conditions, for example, changes in the production and consumption of electrical energy. The nature of the reaction is determined by the inertia of the instruments and the rate of transient processes, i.e., by the dynamic properties. The dynamic properties of thermal electric accumulators are determined by dynamic characteristics. That is, by assessing the change in the temperature of the heated water over time. Thus, the thermal electric accumulator is a dynamic system. Therefore, ensuring the operation of a thermal electric accumulator should be part of such a technological system, which is based on a dynamic system. Representing the design of a technological system as an organization of a complex system, we expand it by building a dynamic subsystem on its foundation – nodes that predict the components of the technological process: charging, discharging, and assessment of functional efficiency. Based on the system-structural substantiation of the technological system architectures the connections of the dynamic subsystem with other units of the technological system based on mathematical modeling of their logical connections that change over time will also be established. This approach enables setting new properties of the dynamic subsystem and nodes of the technological system. Based on the structured systems substantiation of the technological system architectures, the category of relations is considered to be organizing interactions within the elements of the dynamic subsystem and nodes of the technological system. This enables monitoring of the performance, identifying the state of the dynamic subsystem, and confirming new modes of operation from the nodes of the technological system based on the developed cause-and-effect graph method.

Based on the methodological, mathematical, and logical substantiation of the technological systems (Chapter 1), the architecture, mathematical substantiation of the architecture (1), and mathematical substantiation of operational maintenance (2), Smart Grid thermal electric accumulators are proposed (Figure 3.1).

The integrated dynamic subsystem (Figure 3.1) includes the following components: network, thermal electric accumulator, and frequency converter.

When the system design is represented as the organization of a complex system, it is expanded by building up the dynamic subsystem blocks that forecast the process components around its base. Other components of the technological system include the charging, discharging, and functional efficiency estimation units in a coordinated interaction with the dynamic subsystem (Figure 3.1).

The proposed mathematical substantiation of the Smart Grid thermal electric accumulator system architecture (1) (Figure 3.1) is based on the mathematical description of power systems dynamics and the cause-effect graph method (Chapter 1).

Where $TEASSG(\tau)$ – Smart Grid thermal electric accumulator system; τ – time, seconds; $ID(\tau)$ – integrated dynamic subsystem (network, thermal electric accumulator, and frequency converter); $P(\tau)$ – properties of the thermal electric accumulator system components; $x(\tau)$ – impacts (changes in production and consumption of electric energy, etc.; $f(\tau)$ – parameters that are measured: (temperature of the water at the inlet to the thermal electric accumulator and the outlet of the thermal electric accumulator, voltage at the frequency converter input and output); $K(\tau)$ – mathematical coefficients describing the dynamics of water temperature change within the thermal electric accumulator volume; $y(\tau,z)$ – predicted output parameter (water temperature within the thermal electric accumulator volume); z – heater length coordinate, meters; $d(\tau)$ – dynamic parameters (water temperature within the volume); $FI(\tau)$ – the resulting functional information for decision-making; $LC(\tau)$ – logical relations regarding the operability control of the thermal electric accumulator system; $LS(\tau)$ – logical relations regarding the identification of the thermal electric accumulator system state; $R(\tau)$ – logical relations in $TEASSG(\tau)$ to confirm the correctness of decisions made regarding the units of the thermal electric accumulator system. Indices: i – the number of elements of the thermal electric accumulator system; 0,1,2 – initial stationary mode, external, internal nature of impacts.

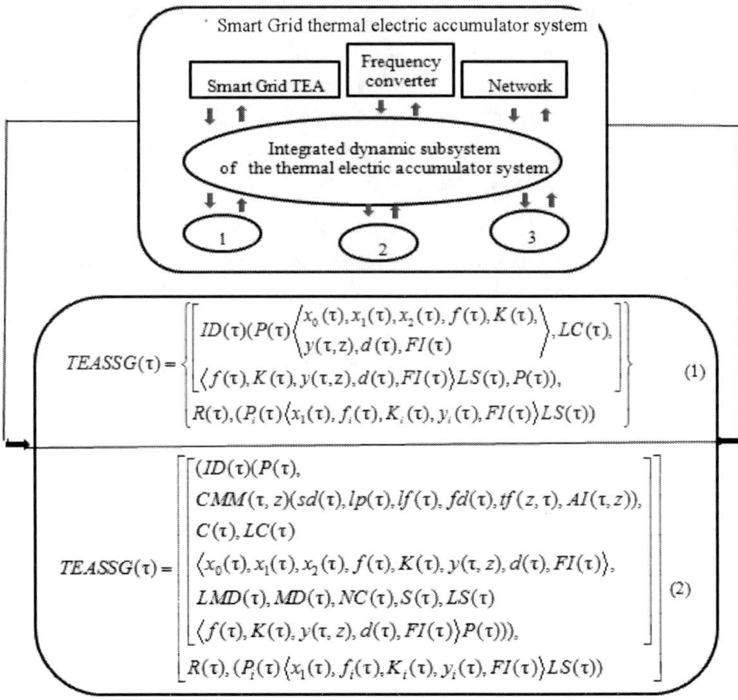

Figure 3.1. Smart Grid thermal electric accumulator system: the architecture: TEA – the thermal electric accumulator; 1 – the charging unit; 2 – the discharging unit; 3 – the functional efficiency assessment unit. Mathematical substantiation of the architecture (1). Mathematical substantiation of maintenance of the operation (2).

The proposed mathematical substantiation of operational maintenance of the Smart Grid thermal electric accumulator system (2) (Figure 3.1) is based on the methodology of the mathematical description of power systems dynamics and the cause-effect graph method (Chapter 1). The basis of the proposed rationale is the mathematical description of the thermal electric accumulator system architecture (1) (Figure 3.1). Prediction of changes in the thermal electric accumulator capacity enables the making of decisions in advance to change the level of power in the thermal electric accumulator.

The mathematical substantiation of the Smart Grid's operational maintenance of the thermal electric accumulator (2) is proposed (Figure 3.1).

Where *TEASSG* (τ) – the operational maintenance of the Smart Grid thermal electric accumulator system; τ – time, seconds; *ID*(τ) – integrated dynamic subsystem (network, thermal electric accumulator, and frequency converter). *P*(τ) – properties of the elements of the integrated dynamic subsystem, units of the thermal electric accumulator system; *CMM*(τ,*z*) – complex mathematical modeling of the dynamics of the water temperature changes within the volume of the thermal electric accumulator system; *sd*(τ) – the input data (the thermal electric accumulator and its type and capacity; *lp*(τ) – the boundary change in parameters (the water temperature within the thermal electric accumulator volume); *lf*(τ) – the levels of operation of the thermal electric accumulator; *fd* (τ) – the obtained parameters (mode parameters of the thermal electric accumulator); *tf*(τ,*z*) – the transfer function of predicted parameters – water temperature within the thermal electric accumulator system volume; *AI*(τ,*z*) – the standard information regarding the evaluation of the maximum permissible change in the water temperature within the thermal electric accumulator system volume; *C*(τ) – the operability control of the thermal electric accumulator system; *LC*(τ) – the logical relations of the thermal electric accumulator system operability control; *x*(τ) – impacts (changes in production and consumption of electric energy, etc.; *f*(τ) – the measured parameters: (the water temperature at the inlet and outlet of the thermal electric accumulator, temperature of the heating element); *K*(τ) – the coefficients of the mathematical description of the change dynamics in the water's temperature within the thermal electric accumulator volume; *y*(τ,*z*) – the output parameter (the water temperature within the thermal electric accumulator volume); *z* – the length coordinate of the heater element, meters; *d*(τ) – the dynamic parameters of estimated water temperature change within the thermal electric accumulator volume; *FI*(τ) – the resulting functional information on decision-making; *LMD*(τ) – the logical relations of decision-making; *MD*(τ) – decision-making; *NC*(τ) – the new conditions of the thermal electric accumulator system operation; $S(\tau)$ – the identification of the thermal electric accumulator system state; $LS(\tau)$ – the logical relations of identification of the thermal electric accumulator system state; $R(\tau)$ – the logical relations between the dynamic subsystem and the charging, discharging, and functional efficiency estimation units included in the thermal electric accumulator system. Indices: i – the number of elements of $TEASSG(\tau)$; 0,1,2 – the initial, external, and internal character of influences.

Mathematical substantiation of the Smart Grid's thermal electric accumulator system architecture (1) and mathematical substantiation of operational maintenance of the Smart Grid thermal electric accumulator system (2) (Figure 2.1) enable maintaining the operation of the thermal electric accumulator system using the following actions:

- operability control ($C(\tau)$) of the dynamic subsystem based on complex mathematical ($CMM(\tau,z)$) and logical ($LC(\tau)$) modeling aimed at obtaining standard ($AI(\tau,z)$) estimate of a change in the water temperature within the thermal electric accumulator system's volume;
- operability control ($C(\tau)$) of the dynamic system based on complex mathematical ($CMM(\tau,z)$) and logical ($LC(\tau)$) modeling aimed at obtaining functional ($FI(\tau)$) estimate of a change in the water temperature within the thermal electric accumulator system's volume;
- decision-making ($MD(\tau)$) with the use of the resulting functional information ($FI(\tau)$) obtained based on logical modeling ($LMD(\tau)$); decision-making to change the power level of the thermal electric accumulator system;
- identification ($S(\tau)$) of the new operating conditions of the thermal electric accumulator system; ($NC(\tau)$) based on logical modeling ($LS(\tau)$) as a part of the dynamic subsystem and confirmation of new operating conditions based on logical modeling ($R(\tau)$) from the units of the thermal electric accumulator system;

Maintaining the Thermal Electric Accumulator Capacity Based on the Predicted Water Temperature Changes

The basis for coordinating the processes of heat and mass transfer during the charging of a thermal electric accumulator is the predicted water temperature changes within the thermal electric accumulator volume when the water temperature is measured at the inlet and outlet of the thermal accumulator. For this, a mathematical model of water temperature change dynamics within the thermal electric accumulator volume during charging is proposed.

According to formulas (1) and (2), a transfer function of the "water temperature – consumption of water within the thermal electric accumulator volume" relation, which enables the prediction of a change in the water temperature, both over time and along the length of the heating element, is represented in the following way:

$$W_{t-G_1} = \frac{K_w \varepsilon \left(1 - L_{out}^*\right)}{(T_w + 1)\beta - 1}\left(1 - e^{-\gamma\xi}\right), \tag{3}$$

where

$$K_w = \frac{m(\theta_0 - t)}{G_0};$$

$$\varepsilon = \frac{\alpha_0 h_{in\,0}}{\alpha_0 h_{out\,0}}$$

$$L_{out}^* = \frac{1}{L_{out} + 1}; \quad L_{out} = \frac{G_w C_w}{\alpha_0 h_{out\,0}} \quad \gamma = \frac{(T_w S + 1)\beta - 1}{\beta};$$

$$\xi = \frac{z}{L_w}; \quad T_w = \frac{g_w C_w}{\alpha_0 h_{out\,0}}$$

$$\beta = T_m S + \varepsilon^* + 1; \quad T_m = \frac{g_m C_m}{\alpha_0 h_{out\,0}} \quad \varepsilon^* = \varepsilon(1 - L_{out}^*);$$

where α is the coefficient of heat emission, kW/(m^2·K); C is the specific heat capacity, kJ/(kg·K); G is the consumption of substance, kg/s; g is the specific mass of substance, kg/m; h is the specific surface, m^2/m; t, θ is the water temperature in the heating element wall, K; z is the length coordinate of the heating element, meters; T_w and T_m are the constants of time characterizing the thermal accumulation ability of water and metal, seconds; m is the dependence indicator of heat emission coefficient on consumption; ι is time, seconds; S is the Laplace transform parameter; $S = \omega j$; ω is frequency, 1/s. Indices: 0 – initial stationary mode; 1 – input to the accumulator; in – input; out – output; w – heated water; m – the metal wall.

The transfer function for the "water temperature – water consumption within the thermal electric accumulator volume" relation (3) is obtained on the basis of solving the system of nonlinear differential equations using the Laplace transform. The system of differential equations includes the state equation as an estimate of the physical model of the accumulator, the charge energy equation, and the heat balance equation for the wall of the built-in heating element in the accumulator. The charging energy equation was developed with the representation of changes in the water temperature not only over time but also along the spatial coordinates of the built-in heating element of the accumulator. The charging energy equation includes the K_w coefficient, which estimates a change in the temperature of the built-in heating element of the accumulator.

A valid part of the transfer function (3) was singled out to estimate a change in the water temperature within the thermal electric accumulator volume:

$$O(\omega)=\frac{(L_1A_1)+(M_{1.}B_1)K_w\varepsilon(1-L_{\text{out}}^*)}{(A_1^2+B_1^2)}. \tag{4}$$

K_w coefficient includes the temperature of the dividing wall θ:

$$\theta=(\alpha((t_1+t_2))/2)+A(t_3+t_4)/2)/(\alpha+A), \tag{5}$$

where t_1 and t_2 are water temperatures at the inlet and outlet to the thermal electric accumulator, K, respectively; t_3 and t_4 are the heating element temperature at the inlet and outlet to the thermal electric accumulator, K, respectively.

$$A = 1/(\delta_m / \lambda_m + 1/\alpha), \tag{6}$$

where δ is the thickness of the heat element wall in the thermal electric accumulator, meters; α is the heat emission coefficient, kW/(m^2·K); λ is the heat conductivity of metal in the thermal electric accumulator element, kW/(m·K). Indices: m – metal wall.

To use the valid part of $O(\omega)$, the following coefficients were obtained:

$$A_1 = \varepsilon^* - T_w T_m \omega^2; \;\; A_2 = \varepsilon^* + 1; \;\; B_1 = T_w \varepsilon\omega + Tw\omega + T_m\omega; \tag{7}$$

$$B_2 = T_m\omega; \;\; C_1 = \frac{A_1 A_2 + B_1 B_2}{A_2^2 + B_2^2}; \;\; D_1 = \frac{A_2 B_1 - A_1 B_2}{A_2^2 + B_2^2}; \tag{8}$$

$$L_1 = 1 - e^{-\zeta C_1}\cos(-\xi D_1); \;\; M_1 = -e^{-\zeta C_1}\sin(-\xi D_1). \tag{9}$$

The implementation of a transfer function (3) obtained by using the operator method for solving the system of nonlinear differential equations, holds a Laplace transform parameter – $S(S = \omega j)$, where ω is the frequency, 1/s. The valid part (4), obtained as a result of mathematical processing of the transfer function, was chosen for the transition from the frequency domain to the time domain in order to obtain the dynamic characteristics of the temperature of the heated water. This part is included in the integral (10), which enables obtaining the estimated change in the tank charge over time using the inverse Fourier transform.

Using the integral of transition from the frequency region to the time region, a change in the water temperature within the thermal electric accumulator volume was determined in the following way:

$$t(\tau, z) = \frac{1}{2\pi}\int_0^\infty K_w(\tau) O(\omega) \sin(\tau\omega / \omega) d\omega, \tag{10}$$

where t is the water temperature within the thermal electric accumulator volume, K.

Thus, a block diagram is proposed for obtaining a reference estimation of water temperature change within the thermal electric accumulator volume, (Figure 3.2) with, for example, initial data for a thermal electric accumulator with a power of 5 kW.

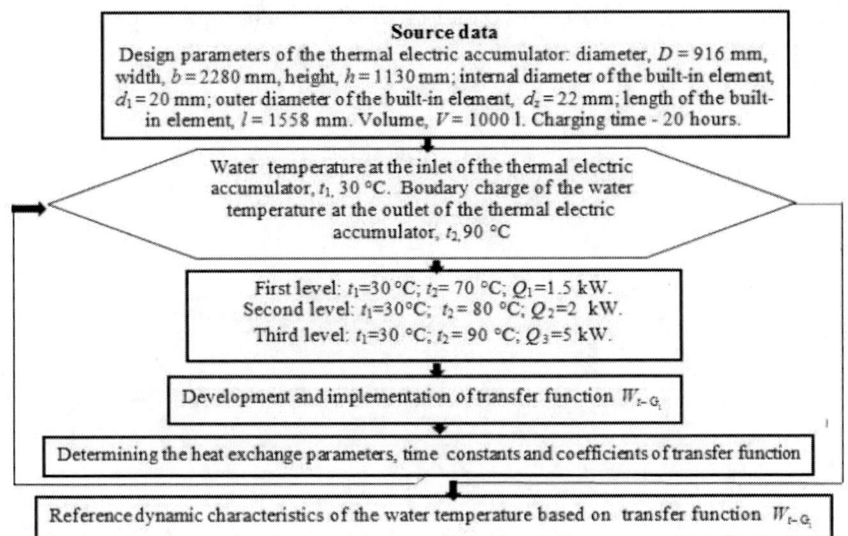

Figure 3.2. Block diagram of comprehensive mathematical modeling of the thermal electric accumulator: Q_1 Q_2, Q_3 – the thermoelectric accumulator power, kW.

Table 3.1. Mode parameters, heat exchange parameters, time constants, and coefficients of the mathematical model of dynamics of the thermal electric accumulator

Functioning levels	N_T, kW	t,°C	G_w, kg/s	α_w, W/(m²·K)	τ_{ch}, hrs.	T_w, s	T_m, s	L_w, m	L_w^*	ε^*	ζ
Level 1	1.5	70	0.0055	774.15	12.3	29.0	2.55	0.43	0,70	0.29	3.62
Level 2	2	80	0.0063	823.2	0.35	27.1	2.39	0.46	0,68	0.36	3.39
Level 3	5	90	0.014	867.9	1.25	25.5	2.28	0.98	0.5	0.45	1.59

Note: N_T – the thermoelectric battery power, kW; t – temperature of local water at the thermoelectric battery outlet, °C; G_w – local water flow rate, kg/s; α_w – coefficient of convective heat transfer from an electric heater element to local water, W/(m²·K); τ_{ch} – thermoelectric battery charging duration.

The following levels of thermal electric accumulator operation have been established for the change in the power: first level – 1.5 kW; second level – 2 kW; third level – 5 kW. They correspond to changes in the water temperature in the outlet of the thermal electric accumulator: 70°C, 80°C, and 90°C with the water temperature at the inlet of the thermal electric accumulator at 30°C and the heated water consumption of 0.0055 kg/s, 0.0063 kg/s, and 0.014 kg/s respectively. According to the formulas (1)–(3) and the proposed block diagram (Figure 3.2), the results of reference information obtained based on complex mathematical modeling of the Smart Grid thermal electric accumulator are presented (Table 3.1).

Time constants and the coefficients that are components of the mathematical dynamics model (3) presented in Table 3.1 were obtained based on the mode and heat exchange parameters.

Based on the proposed mathematical substantiation of operational maintenance of the Smart Grid thermal electric accumulator system (1) to (3), the block diagram for control of serviceability of the thermal electric accumulator system (Figure 3.3) has been developed.

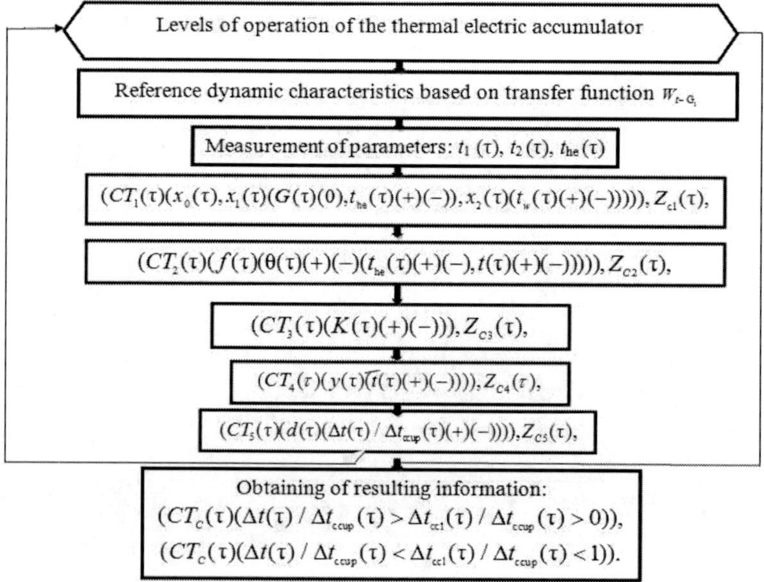

Figure 3.3. Block diagram of the thermal electric accumulator system operational control: t_1, t_2, t_{he} – local water temperature at the inlet and outlet of the thermoelectric battery, the temperature of the heating element, respectively, °C; CT – event control; Z – logical relations; d – dynamic parameters; x – effects; f – parameters measured; y – parameters predicted; K – coefficients of mathematical description; ι – time. Indices: c – operability control; ccup – the constant calculated value of the upper operational level parameter; ccl – the constant calculated value of the operational level parameter; 0,1,2 – initial stationary mode, external, internal influences; 3 – dynamics equations coefficients; 4 – significant predicted parameters; 5 – dynamic parameters.

Operability control of the thermal electric accumulator system (Figure 3.3) enables obtaining the resulting information for advance decision-making about the maintenance of charging and discharging the thermal electric accumulator. Based on the proposed mathematical substantiation (1) to (3), the block diagram of operational maintenance of operation of the Smart Grid thermal electric accumulator system (Figure 3.4) has been developed.

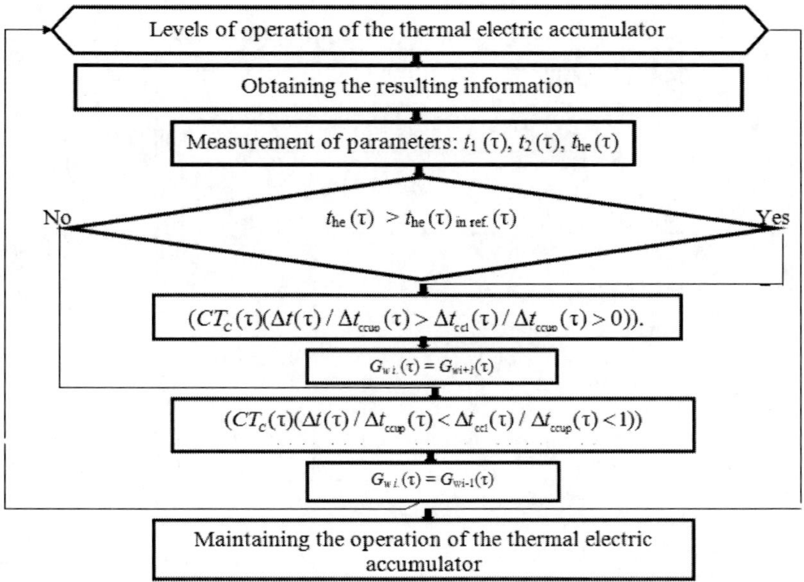

Figure 3.4. Block diagram of operational maintenance of the thermal electric accumulator system: t_1, t_2, t_{he} – local water temperature at the inlet and outlet to the thermoelectric battery, the heating element temperature, respectively, °C; G_w – local water flow rate, kg/s; ι – time, seconds. Indices: i – number of operational levels; in ref – the parameter reference input value; ccup – constant calculated value of the upper operational level parameter; ccl – constant calculated value of the operational level parameter.

Results and Discussion

The Smart Grid System of Maintaining the Operation of the Thermal Electric Accumulator at the Decision-Making Level

An integrated system is designed to support a change in the capacity of the thermal electric accumulator (Table 3.2, Figure 3.5) based on predicting the water change in temperature within the thermal electric accumulator volume through continuous measurement of the water temperature at the inlet and outlet of the thermal electric accumulator.

Table 3.2. Integrated system for supporting the functioning of the thermal electric accumulator system

Time, τ, 10^3 s	Change in parameters	$\Delta t(\tau)/\Delta t_1(\tau)$	$t(\tau)$,°C
0	Charge – discharge N_T = 1.5 kW; t_1 = 30°C; t_2 = 70°C; t_3 = 75°C; t_4 = 70°C; G_w = 0.0055 kg/s	0	30
3	Charge – discharge N_T = 1.5 kW; t_1 = 30°C; t_2 = 70°C; t_3 = 75°C; t_4 = 70°C; G_w = 0.0055 kg/s	0.0597	33.58
6	Charge – discharge N_T = 1.5 kW; t_1 = 30°C; t_2 = 70°C; t_3 = 75°C; t_4 = 70°C; G_w = 0.0055 kg/s	0.1977	41.86
9	Charge – discharge N_T = 1.5 kW; t_1 = 30°C; t_2 = 70°C; t_3 = 75°C; t_4 = 70°C; G_w = 0.0055 kg/s	0.3633	51.80
12	Charge – discharge N_T = 1.5 kW; t_1 = 30°C; t_2 = 70°C; t_3 = 75°C; t_4 = 70°C; G_w = 0.0055 kg/s	0.5175	61.05
15	Decision making N_T = 2 kW; t_1 = 30°C; t_2 = 80°C; t_3 = 85°C; t_4 = 80°C G_w = 0.0063 kg/s	0.7000	72
18	Decision-making N_T = 5 kW; t_1 = 30°C; t_2 = 90°C; t_3 = 95°C; t_4 = 90°C G_w = 0.014 kg/s	0.8872	83.23
21	Charge – discharge N_T = 5 kW; t_1 = 30°C; t_2 = 90°C; t_3 = 95°C; t_4 = 90°C; G_w = 0.014 kg/s	0.9717	87.70
24	Charge – discharge N_T = 5 kW; t_1 = 30°C; t_2 = 90°C; t_3 = 95°C; t_4 = 90°C; G_w = 0.014 kg/s	0.9900	88.80
27	Charge – discharge N_T = 5 kW; t_1 = 30°C; t_2 = 90°C; t_3 = 95°C; t_4 = 90°C; G_w = 0.014 kg/s	1	90
30	Charge – discharge N_T = 5 kW; t_1 = 30°C; t_2 = 90°C; t_3 = 95°C; t_4 = 90°C; G_w = 0.014 kg/s	1	90

Note: N_T – thermoelectric battery power, kW; t_1, t_2, t_3, t_4 – water temperature at the inlet and outlet of the thermoelectric battery, the temperature of the heating element at the inlet and outlet of the thermoelectric battery, respectively,°C; G_w – local water flow rate, kg/s; ι – time, seconds. Index: 1 – the constant calculated value of the upper operational level parameter.

The water temperature within the thermoelectric battery volume at a specified point in time was determined as follows:

$$t_{i+1}(\tau) = t_i + \\ + \begin{pmatrix} \Delta t_{i+1}(\tau)/\Delta t_1(\tau) - \\ -\Delta t_i(\tau)/\Delta CE_1(\tau) \end{pmatrix} (t_2 - t_1), \quad (11)$$

where t – water temperature within the thermoelectric battery volume, °C; t_1, t_2 – initial and final values of the water temperature, °C; τ – time, seconds. Indices: 1 – the constant calculated value of the upper operational level parameter; i – the number of levels of operation of the thermoelectric battery.

For example, over a period of $15 \cdot 10^3$ s (4.17 hrs), an increase in the water temperature within the thermal electric accumulator volume was predicted to reach the level of 72°C at a water flow rate of 0.0055 kg/s. The water temperature was determined by formula (18) as follows (Table 3.2, Figure 3.5):

72°C = 61.05+ (0.7–0.5175) (90–30).

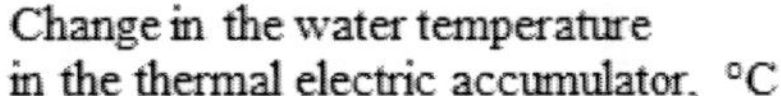

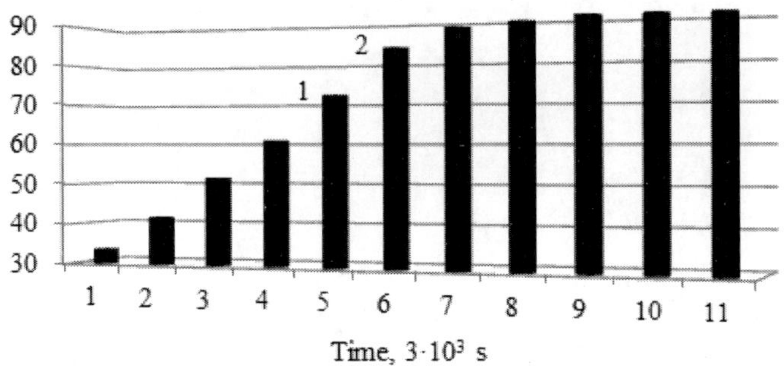

Figure 3.5. Operational maintenance of the thermal electric accumulator. 1 and 2 are the points where decisions were made on the change in the local water flow rate level.

In this period of time, it is necessary to make a preliminary decision to increase the power of the thermal electric accumulator to the level of 2 kW and increase the local water flow to the level of 0.0063 kg/s.

The use of an integrated system for estimating the change in the temperature of heated water within the accumulator volume, obtained based on matching the thermal and mass-exchange processes of discharging and charging, enables timely decisions to be made on changing the power of the thermal electric accumulator by altering the local water flow rate, which enables reducing the charging time by up to 30%.

For example, this technology is used in the developed integrated Smart Grid system to support the operation of a wind-solar power plant by predicting changes in the battery capacity (Chapter 6).

Changes in the speed of rotation of the circulation pump's electric motor are given for the conditions of the flow rate and of heated water temperature changes, thus reducing the charging duration by up to 30%.

The storage battery and the thermoelectric accumulator, as parts of the solar power system network (Chapter 4), acquire the additional status of voltage regulators in the distribution system.

Preliminary decisions are made to change the capacitance of the thermoelectric accumulator when redistributing the accumulated electrical energy.

Changing the level of transmission of electrical energy to the grid enables maintaining voltage within the distribution system by maintaining the power factor of the solar power plant grid. The peak load on the power system is prevented, which reduces the consumption of electricity from the network by up to 14%.

Chapter 4

Smart Grid Technology for Maintaining the Operation of a Solar Electric System Network

Abstract

It is proposed to maintain the voltage within the distribution system by forecasting the change in the storage battery capacity in order to make anticipatory decisions on the change in the power of the thermoelectric accumulator, in relation to the maintenance of the power factor of the solar electric system network. The change in the ratio of the voltage at the frequency converter output and the voltage within the distribution system is assessed by measuring the voltage at the input of the hybrid inverter. A comprehensive system has been developed to support the operation of a solar power plant network based on predicting changes in battery capacity and power factor. The accumulator battery and the thermoelectric accumulator as part of the solar electric system network acquire the additional status of voltage regulators in the distribution system. Preliminary decisions are made to change the capacity of a thermoelectric accumulator to redistribute the accumulated electrical energy. Changing the level of transmission of electrical energy to the grid enables maintaining the voltage in the distribution system by maintaining the power factor of the solar power plant grid. Continuous measurement of the voltage at the hybrid inverter input, at the frequency converter output, and within the distribution network is performed. The change in the ratio of voltages at the frequency converter output and within the distribution network is estimated. Peak load on the power system is prevented, and reducing power consumption from the grid by up to 14%.

Keywords: solar electric system, rechargeable battery, thermoelectric battery, maintaining the voltage within the distribution system

Introduction

The distributed generation of electricity using renewable sources requires intelligent systems for managing electricity flows and consumption. Smart

Grid technologies, demand management systems and energy storage are new components for the integration of distributed energy generation into the energy system.

In terms of connection to intelligent control systems, the author (Chapter 2) proposes to predict voltage changes by measuring the temperature of the electrolyte within the battery volume. An energy-saving technology has been developed for the operation of a storage battery; it prevents overcharging and stops the discharge based on the coordination of electrochemical and diffusion processes of discharging and charging. Integrated Smart Grid Systems for harmonization of production and consumption of electric power based on the predicted changes in the battery capacity were developed (Chapter 5). Advance decisions on the change in power transmission capacity enabled regulating the voltage within the distribution system by maintaining the power factor of the photoelectric charging station. Voltages at the hybrid inverter input and within the distribution system were measured so as to assess their ratio.

An urgent task regarding the further development of Smart Grid technologies is the maintenance of the power factor of the solar electric system network based on the integrated management of the accumulation of electric energy and heat. It is known that the thermoelectric accumulator is controlled according to the thermostat principle. That is, when the required local heated water temperature is established, the thermoelectric accumulator is disconnected from the power supply. Not using the change in the local water flow rate during the thermoelectric accumulator charging period increases the charging duration and leads to significant consumption of electrical energy. For the purpose of comprehensive management of electrical energy and heat accumulation, it is necessary to predict changes in the battery capacity. Adopting anticipatory decisions to change the power of the thermoelectric accumulator enables maintaining the voltage within the distribution system. The maintenance of the power factor within the solar electric system grid is performed by measuring the voltage at the hybrid inverter input, the frequency converter output, and within the distribution system. The change in the ratio of the voltage at the frequency converter output and the voltage within the distribution system is evaluated. Maintaining the power factor of the solar electric system network when making anticipatory decisions on changing the power of electric energy transmission to the network enables preventing the peak load on the electric system when satisfying consumer demand. The optimization of distributed electrical energy generation usually uses the improvement of intelligent control systems for both the production and consumption of electrical energy. Thus, work (Yhosvany Soler-Castillo, Julio César Rimada, Luis Hernández, Gema Martínez-Criado, 2021), which estimates energy losses to the photovoltaic module, is devoted to forecasting the efficiency of electric energy production. The estimate of the cost of energy based on the discounting method cannot be generalized. It is not possible to generalize and distribute the economic scheduling algorithm for minimizing the total costs of production (Zhiyong Li, Gang Chen, 2022), because it limits the balance of demand and the volume of electric energy production. The work (Nishant Jha, Deepak Prashar, Mamoon Rashid, Sachin Kumar Gupta, R. K. Saket, 2021) is devoted to early forecasting of electric load efficiency, in which a neural model of electric energy planning and distribution is proposed, but without coordination with production. The paper (Saad, A. A., Faddel, S., Mohammed, O., 2019) presents the results of the implementation of the stochastic optimization algorithm for distributed generation of electric energy using fuzzy logic. The relationship between the load on the electrical system, operating costs, and the flexibility of managing distributed generation has been established. It is proposed to establish the limit level of electric power generation using the utility network as virtual storage in order to maintain the flexibility of management. However, the design and management strategy, on which the results presented in this work are calculated, do not enable expanding the level of distributed energy generation. Reconciliation of production and consumption of electrical energy presented in the works (Heydar Chamandoust, Abozar Hashemi, Salah Bahramara, 2021, Chunhe Song, Yingying Sun, Guangjie Han, Joel J. P. C., 2021) requires uncertainty models. For example, using the Monte Carlo method (Heydar Chamandoust, Abozar Hashemi, Salah Bahramara, 2021), or genetic algorithm (Chunhe Song, Yingying Sun, Guangjie Han, Joel J. P. C., 2021), which is based on deep learning for long-term memory. Therefore, for example, a blockchain-based data aggregation scheme for preserving privacy within smart networks is proposed (Parminder Singh, Mehedi Masud, M. Shamim Hossain, Avinash Kaur, 2021). Uncertainty can be prevented by using, for example, large-scale seasonal heat storage (ATES) (Rostampour, V., Jaxa-Rozen, M., Bloemendal, M., Kwakkel, J., Keviczky, T. Aquifer, 2019) for intelligent management of distributed energy generation. However, the exchange of information between ATES systems regarding dynamic control does not establish a connection between the use of accumulation and the assessment of the power factor change. Management of the production and consumption of electrical energy usually takes place with the use of additional devices for voltage regulation within the distribution system and requires additional costs. The storage battery and the thermoelectric accumulator become voltage regulation elements within the distribution system while ensuring comprehensive energy consumption and maintaining the power factor of the solar electric system grid. Forecasting the change in battery capacity provides an opportunity to make anticipatory decisions on changing the thermoelectric accumulator capacity. There is a change in the rotation speed of the electric motor of the circulation pump in relation to the change in the flow rate and temperature of the heated water. Therefore, it is proposed to measure the voltage at the hybrid inverter input, the voltage at the frequency converter output, and the voltage and frequency within the distribution system. The ratio of the voltage at the frequency converter output and the voltage within the distribution system is estimated. Making anticipatory decisions to change the level of electric energy transmission to the network enables adjustment of the voltage within the distribution system to match energy production and consumption, thus preventing the peak load on the energy system.

Methodological and Mathematical Substantiation

Based on the methodological, mathematical, and logical substantiation of the technological systems (Chapter 1) the architecture, mathematical substantiation of the architecture (1), and mathematical substantiation of operational maintenance (2) of the Smart Grid solar electric system network is proposed (Figure 4.1).

The solar electric system network is a dynamic system, the operation of which is the reproduction of a change in external, and internal influences and initial conditions. For example, changes in solar radiation, change in electrical energy and heat consumption, etc. Therefore, when designing a solar electric system network, an integrated dynamic subsystem is laid down as its foundation (Figure 4.1). The integrated dynamic subsystem includes the following components: the power grid, a photovoltaic module, a hybrid inverter, a rechargeable battery, a two-way Smart Meter counter for changing the level of electric energy transmission to the network; a two-section storage tank, the upper section of the two-section storage tank, frequency converter. When representing the system design as the organization of a complex system, it was expanded by building up the dynamic subsystem blocks that forecast the process components around its base. Other components of the technological system include the units for charging, discharging, and functional efficiency estimation in a coordinated interaction with the dynamic subsystem (Figure 4.1).

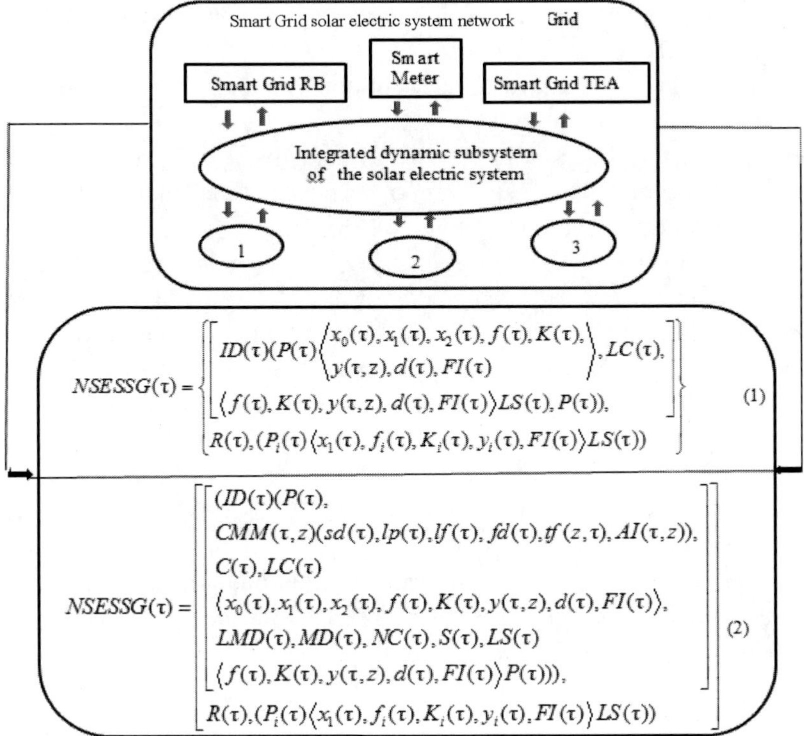

Figure 4.1. Smart Grid solar electric system network: the architecture: RB – rechargeable battery; Smart Meter is a two-way counter of changes in the level of power transmission to the network; TEA – thermoelectric accumulator; 1 – the charging unit; 2 – the discharging unit; 3 – the unit of assessing the functional efficiency. Mathematical substantiation of the architecture (1). Mathematical substantiation of operational maintenance (2).

The mathematical substantiation of the architecture of the Smart Grid solar electric system network (1), (Figure 4.1), based on the mathematical description of the dynamics of power systems, the cause-effect graph method (Chapter 1) is proposed.

Where $NSESSG(\tau)$ – Smart Grid solar electric system network; τ – time, seconds; $ID(\tau)$ – integrated dynamic subsystem (power grid, photovoltaic module, hybrid inverter, rechargeable battery, two-way Smart Meter counter for changing the level of electric energy transmission to the network; two-section storage tank, upper section of the two-section storage tank, frequency converter); $P(\tau)$ – properties of the components of the solar electric system network; $x(\tau)$ – impacts (change in solar radiation, change in consumption of electrical energy and heat, etc; $f(\tau)$ – parameters that are measured: (voltage at the hybrid inverter input, voltage at the frequency converter output, voltage frequency, voltage in the distribution system); $K(\tau)$ – coefficients of the mathematical description of the dynamics of the battery capacity change, the power factor of the solar electric system network; $y(\tau,z)$ – predicted output parameters (battery capacity, power factor of the solar electric system grid); z – coordinate of the length of the battery plates, meters; $d(\tau)$ – dynamic parameters of battery capacity change, power factor of the solar electric system network; $FI(\tau)$ – the resulting functional information for decision-making; $LC(\tau)$ – logical relations regarding the control of the solar electric system network operability; $LS(\tau)$ – logical relations regarding the identification of the state of the solar electric system network; $R(\tau)$ – logical relations in $WSESSG$ (τ) to confirm the correctness of decisions made from the units of the solar electric system network. Indices: i – the number of elements of the solar electric system network; 0,1,2 – initial stationary mode, external, internal nature of impacts.

The proposed mathematical substantiation of operational maintenance of the solar electric system network (2), (Figure 4.1) is based on the methodology of the mathematical description of the dynamics of power systems, the cause-effect graph method (Chapter 1). The basis of the proposed rationale is the mathematical description of the architecture of the solar electric system network (Figure 4.1).

Where $NSESSG(\tau)$ – operational maintenance of the Smart Grid solar electric system network; τ–time, seconds; $ID(\tau)$ – integrated dynamic subsystem (power grid, photovoltaic module, hybrid inverter, rechargeable battery, two-way Smart Meter counter for changing the level of electric energy transmission to the network; two-section storage tank, upper section of the two-section storage tank, frequency converter). $P(\tau)$ – the properties of the elements of the integrated dynamic subsystem, units of the solar electric system network; complex mathematical modelling of the dynamics of changes in battery capacity, the power of the upper section of the two-section storage tank, the power factor of the solar electric system grid; $CMM(\tau, z)$ – complex mathematical modelling of the dynamics of changes in battery capacity, the power of the upper section of the two-section storage tank, the power factor of the solar electric system grid; $sd(\tau)$ – the input data (the power of photoelectric module, the power of thermoelectric accumulator, the rechargeable battery and its type and capacity, the two-way Smart Meter counter and its type; $lp(\tau)$ – the boundary change in parameters (the voltage at the hybrid inverter input and at the frequency converter output, as well as within the distribution system; $lf(\tau)$ – the levels of operation of the solar electric system grid; $fd\ (\tau)$ – the obtained parameters (mode parameters of the solar electric system grid); $tf(\tau,z)$ – the transfer function of predicted parameters – the battery capacity, power factor of the solar electric system grid; $AI(\tau, z)$ – reference information regarding the assessment of the change in the capacity of the storage battery, the power factor of the solar electric system grid, $C(\tau)$ – control of the operational capacity of the integrated dynamic subsystem, $MD(\tau)$ – making decisions on the change of the capacity of the upper section of the two-section tank - storage; $S(\tau)$ – identification of the state of the electrical system; $LC(\tau)$, $LMD(\tau)$, $LS(\tau)$ – logical relations in $C(\tau)$, $MD(\tau)$, $S(\tau)$, respectively; $FI(\tau)$ is the resulting functional information; $NC(\tau)$ – new operating conditions as a result of decision-making; $x(\tau)$ – impacts (change in solar radiation, change in consumption of electrical energy and heat, etc); $f(\tau)$ – measured parameters (voltage at the hybrid inverter input, voltage at the frequency converter output, voltage frequency, voltage in the distribution system); $K(\tau)$ – coefficients of the mathematical description of the dynamics of the battery capacity change, the power factor of the solar electric system grid); $y(\tau,z)$ – output parameters (battery capacity, power factor of grid solar electric system); $d(\tau)$ – dynamic parameters of battery capacity change, power factor of the solar electric system network; z – coordinate of the length of the battery plates, meters; τ – time, seconds. Indices: i – components of $NSESSG(\tau)$; 0,1,2 – initial mode, external, internal nature of influences.

Mathematical substantiation of the architecture of the Smart Grid solar electric system network (1) and mathematical substantiation of operational maintenance of the Smart Grid solar electric system network (2) (Figure 4.1) enables maintaining the operation of the solar electric system network using the following actions:

- operability control ($C(\tau)$) of the dynamic subsystem based on complex mathematical ($CMM(\tau,z)$) and logical ($LC(\tau)$) modelling regarding obtaining standard ($AI(\tau,z)$) estimate of a change in the battery capacity, power factor of the solar electric system network;
- operability control ($C(\tau)$) of the dynamic system based on complex mathematical ($CMM(\tau,z)$) and logical ($LC(\tau)$) modelling regarding

the obtained functional ($FI(\tau)$) estimate of a change in the battery capacity, power factor of the solar electric system network;

- decision-making ($MD(\tau)$) with the use of the functional resulting information (FI (τ)), obtained based on logical modelling ($LMD(\tau)$); decision-making to change the level of power transmission to the network to maintain the power factor of the solar electric system network;
- identification ($S(\tau)$) of the new operational conditions of the solar electric system network ($NC(\tau)$) based on logical modelling ($LS(\tau)$) as a part of the dynamic subsystem and confirmation of new operating conditions based on logical modelling ($R(\tau)$) from the units of the solar electric system.

Maintaining the Voltage within the Distribution System Based on a Prediction of Changes in the Battery Capacity

According to formulas (1) and (2) (Figure 4.1), it is proposed to forecast changes in the battery capacity and the power factor of the solar electric system grid. The voltage at the hybrid inverter input, the voltage at the frequency converter output and the voltage and frequency within the distribution system are measured. The change in the ratio of voltage at the frequency converter output and voltage in the distribution system is evaluated. Transfer functions for the following relations: "battery capacity – voltage at the frequency converter output," and "power factor of the solar electric system network – voltage within the distribution system" are presented below:

$$W_{CE-U_2} = \frac{K_{ce}K}{(T_e S+1)\beta\text{-}1}\left(1-e^{-\gamma\xi}\right), \tag{3}$$

$$W_{pf-U_3} = \frac{K_{pf}K}{(T_e S+1)\beta\text{-}1}\left(1-e^{-\gamma\xi}\right), \tag{4}$$

where

$$K_{ce} = \frac{I_1 U_1}{(U_2 - U_3)};\; K_{pf} = \frac{I_1 U_1 - I_2 (U_2 - U_3)}{N};\; K = \frac{m(\theta_0 - \sigma_0)}{G_{e0}};$$

$$L = \frac{G_e C_e}{\alpha_0 h_0};\; T_e = \frac{g_e C_e}{\alpha_0 h_0};\; \beta = T_m S + \varepsilon + 1;\; T_m = \frac{g_m C_m}{\alpha_0 h_0};$$

$$\varepsilon = (1 - L);\; \gamma = \frac{(T_e S + 1)\beta\text{-}1}{\beta};\; \xi = \frac{z}{L},$$

where *CE* is the battery capacity, ampere-hours; *PF* is the power factor of the solar electric system network; I_1, I_2 – currents at the hybrid inverter input, at the frequency converter output, respectively, A; U_1, U_2, U_3 – voltage at the hybrid inverter input, at the frequency converter output, in the distribution system, respectively, volts; *N* is the power of the solar electric system network, kW; *C* – specific heat capacity, kJ/(kg·K); α – heat transfer coefficient, kW/(m^2·K); *G* – substance consumption, kg/s; *g* – specific mass of substance, kg/m; *h* – specific surface, m^2/m; σ, θ – the temperature of the electrolyte at the outlet of the battery and the separating wall, respectively, K; z – coordinate of the length of the battery plates, meters; T_e, and T_m are time constants characterizing the thermal accumulative capacity of the electrolyte, metal, respectively, seconds; m is an indicator of the dependence of the heat transfer coefficient on the consumption; ι – time, seconds; S is the Laplace transform parameter; S = ωj; ω – frequency, 1/s. Indices: 0 – initial stationary mode; ce – capacity; e – electrolyte; m is a metal wall.

Transfer functions for "battery capacity – voltage at the frequency converter output," and "power factor of the solar electric system network – voltage in the distribution system" relations were obtained by solving a system of nonlinear differential equations using the Laplace transform. The systems of differential equations include an equation of state as an estimate of the physical model of the solar electric system network, an equation of energy of the battery charge and discharge, and an equation of heat balance for the wall of the battery plates. The equation of charge and discharge energies was composed with the representation of a change in electrolyte temperature within the plate pores and above the plates both over time and along the spatial coordinate of the battery plates. The transfer functions include coefficients K_{ce} and K_{pf}, which estimate changes in the battery capacity and the power factor of the solar electric system network. When analysing the obtained mathematical model, internal parameters to be diagnosed as being a part of coefficients of the equations of dynamics were established to be K_{ce} and K_{pf}. In real operational conditions of the power system at a transition from stationary states and under external and internal effects, reorganization of the coefficients of dynamic equations in time occurs because of a change in the diagnosed internal parameters.

The valid parts of the transfer functions were singled out:

$$O(\omega)=\frac{(L_1A_1)+(M_1B_1)(1-L)}{(A_1^2+B_1^2)}. \quad (5)$$

The K factor includes the temperature of the separating wall θ:

$$\theta=(\alpha_e(\sigma_1+\sigma_2)/2)+(A(t_1+t_2)/2)/(\alpha_e+A), \quad (6)$$

where σ_1 and σ_2 are electrolyte temperatures at the battery input and output, K, respectively; t_1 and t_2 are electrolyte temperatures within the plate pores and above the plates at the battery input and output, respectively, K; α is the heat transfer factor, kW/(m^2·K). Index: e – electrolyte.

$$A=1/(\delta_m/\lambda_m+1/\alpha), \quad (7)$$

where δ is the battery plate wall thickness, meters; λ is the thermal conductivity of the metal of the battery plate, kW/(m·K). Index m is the metal wall of the battery plate.

To use the valid part $O(\omega)$, the following factors were obtained:

$$A_1=\varepsilon-T_eT_m\omega^2; \quad (8)$$

$$A_2=\varepsilon+1; \quad (9)$$

$$B_1=T_e\varepsilon\omega+T_e\omega+T_m\omega; \quad (10)$$

$$B_2=T_m\omega; \quad (11)$$

$$C_1 = \frac{A_1 A_2 + B_1 B_2}{A_2^{\ 2} + B_2^{\ 2}}; \tag{12}$$

$$D_1 = \frac{A_2 B_1 - A_1 B_2}{A_2^{\ 2} + B_2^{\ 2}}; \tag{13}$$

$$L_1 = 1 - e^{-\zeta C_1} \cos(-\xi D_1); \tag{14}$$

$$M_1 = -e^{-\zeta C_1} \sin(-\xi D_1). \tag{15}$$

The transfer functions (3) and (4) which were obtained based on the use of the operator method of solving the system of nonlinear differential equations, include the Laplace transform parameter – S (S = ωj), where ω is the frequency, 1/s. To switch from frequency to time, a valid part (5), obtained as a result of the mathematical treatment of transfer functions, was singled out. It is this part that is included in the integrals (16) and (17), which enables obtaining dynamic characteristics of a change in the battery capacity, and power factor of the solar electric system network using the inverse Fourier transform:

$$CE(\tau) = \frac{1}{2\pi} \int_0^{\infty} K_{ce} KO(\omega) \sin(\tau\omega/\omega) d\omega, \tag{16}$$

$$PF(\tau) = \frac{1}{2\pi} \int_0^{\infty} K_{pf} KO(\omega) \sin(\tau\omega/\omega) d\omega, \tag{17}$$

where *CE* is the battery capacity, ampere-hours; *PF* is the power factor of the solar electric system network.

Thus, a block diagram (Figure 4.2) is proposed for obtaining of reference estimation of change in the battery capacity and power factor of the solar electric system network, with, for example, the initial data of a solar electric system network with power of 10 kW and a thermal electric accumulator with power of 2.5 kW.

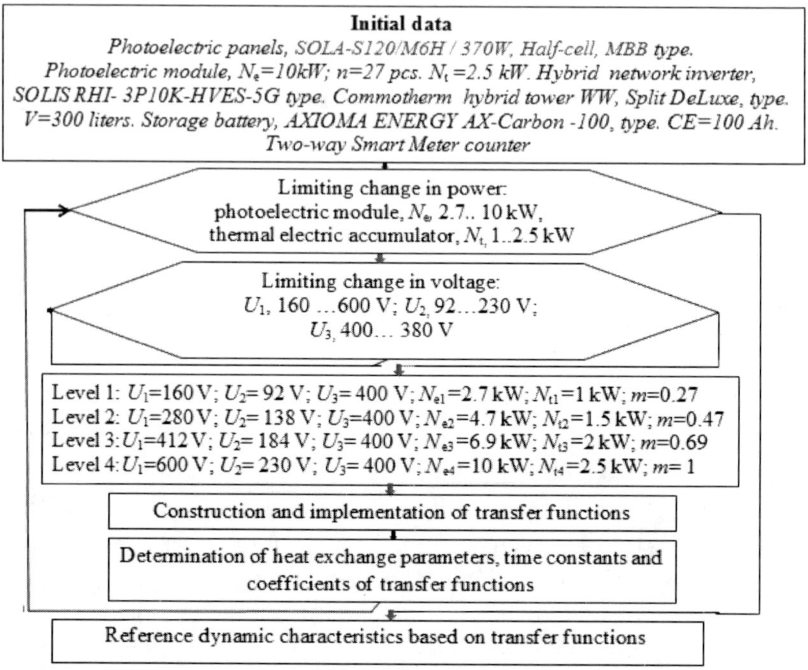

Initial data

Photoelectric panels, SOLA-S120/M6H / 370W, Half-cell, MBB type. Photoelectric module, N_e=10kW; n=27 pcs. N_t=2.5 kW. Hybrid network inverter, SOLIS RHI- 3P10K-HVES-5G type. Commotherm hybrid tower WW, Split DeLuxe, type. V=300 liters. Storage battery, AXIOMA ENERGY AX-Carbon -100, type. CE=100 Ah. Two-way Smart Meter counter

Limiting change in power: photoelectric module, N_e, 2.7.. 10 kW, thermal electric accumulator, N_t, 1..2.5 kW

Limiting change in voltage: U_1, 160 ...600 V; U_2, 92...230 V; U_3, 400... 380 V

Level 1: U_1=160 V; U_2= 92 V; U_3= 400 V; N_{e1}=2.7 kW; N_{t1}=1 kW; m=0.27
Level 2: U_1=280 V; U_2= 138 V; U_3=400 V; N_{e2}=4.7 kW; N_{t2}=1.5 kW; m=0.47
Level 3: U_1=412 V; U_2= 184 V; U_3= 400 V; N_{e3}=6.9 kW; N_{t3}=2 kW; m=0.69
Level 4: U_1=600 V; U_2= 230 V; U_3= 400 V; N_{e4}=10 kW; N_{t4}=2.5 kW; m= 1

Construction and implementation of transfer functions

Determination of heat exchange parameters, time constants and coefficients of transfer functions

Reference dynamic characteristics based on transfer functions

Figure 4.2. Block diagram of comprehensive mathematical modelling of the solar electric system network: N_e, N_t – the power of photoelectric module and thermoelectric accumulator, respectively, kW; *CE* – battery capacity, ampere-hours; U_1, U_2, U_3 – voltage at the hybrid inverter input and at the frequency converter output, as well as within the distribution system, respectively, volts; *n* – the number of photoelectric panels; *m* – the power level of the solar electric system network.

The following levels of operation of the solar electric system network have been established for the change in the voltage at the hybrid inverter input and at the frequency converter output: first level: 160–92V; second level: 280–138V; third level: 412–184V; fifth level: 600–230V. They correspond to changes in the power of the photoelectric module: 2.3 kW, 2.7 kW, 4.7 kW, 6.9 kW, 10 kW, the power of thermoelectric accumulator: 1 kW, 1.5 kW, 2 kW, 2.5 kW, the power level of the solar electric system network: 0.27, 0.47, 0.69, 1, respectively.

According to the formulas (1) – (4) and the proposed block diagram (Figure 4.2), the results of reference information obtained on the basis of complex mathematical modelling of the Smart Grid solar electric system network are presented (Tables 4.1 – 4.3).

Table 4.1. Mode parameters of the solar electric system network

Operation levels	N_e, kW	N_t, kW	t,°C	G, kg/s	τ, hours	U_1, V	U_2, V	U_3, V	m
First level	2.7	1	40	0.007	2	160	92	400	0.27
Second level	4.7	1,5	45	0.009	0.21	280	138	400	0.47
Third level	6.9	2	50	0.011	0.12	412	184	400	0.69
Fourth level	10	2,5	55	0.012	0.077	600	230	400	1

Note: Ne – the power of the solar electric system network, kW; Nt – the power of thermoelectric accumulator, kW; t – local water temperature, °C; G – local water consumption, kg/s; τ – charging time of the thermoelectric accumulator, hours; U1, U2, U3 – voltage at the hybrid inverter input and at the frequency converter output, as well as within the distribution system, respectively, volts; m – the power level of the solar electric system network.

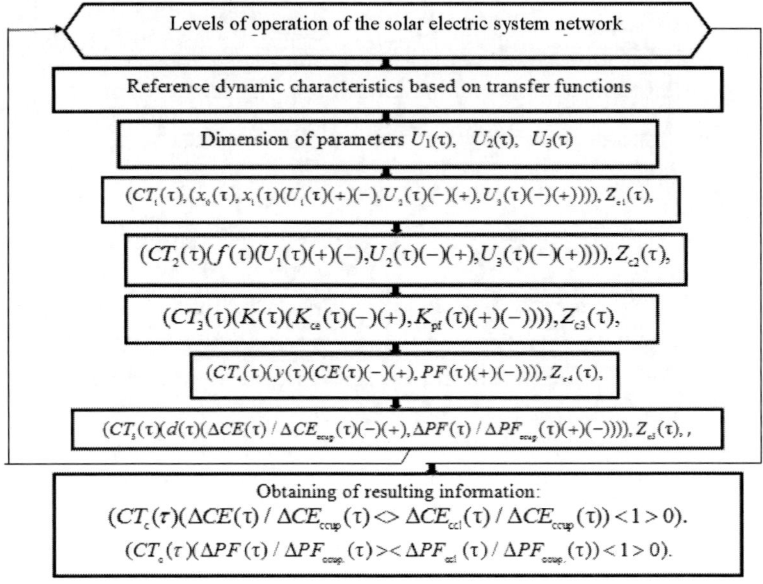

Figure 4.3. Block diagram of operational control of the solar electric system network: U_1, U_2, U_3 – voltage at the hybrid inverter input and at the frequency converter output, as well as within the distribution system, respectively, volts; *CE* – battery capacity, ampere-hours; *KF* – power factor of the solar electric system network; *CT* – event control; *Z* – logical relations; *d* – dynamic parameters; *x* – effects; *f* – parameters measured; *y* – parameters predicted; *K* – coefficients of mathematical description; ι – time. Indices: *c* – control of operability; ccup – the constant calculated value of the parameter of the upper level of operation; ccl – the constant calculated value of the operation level parameter; 0,1,2 – initial stationary mode, external, internal influences; 3 – coefficients of dynamics equations; 4 – significant predicted parameters; 5 – dynamic parameters.

Time constants and the coefficients that are components of mathematical models of dynamics (3) and (4) presented in Table 4.3 were obtained based on the parameters of heat exchange for charging and discharging the battery presented in Tables 4.1 and 4.2.

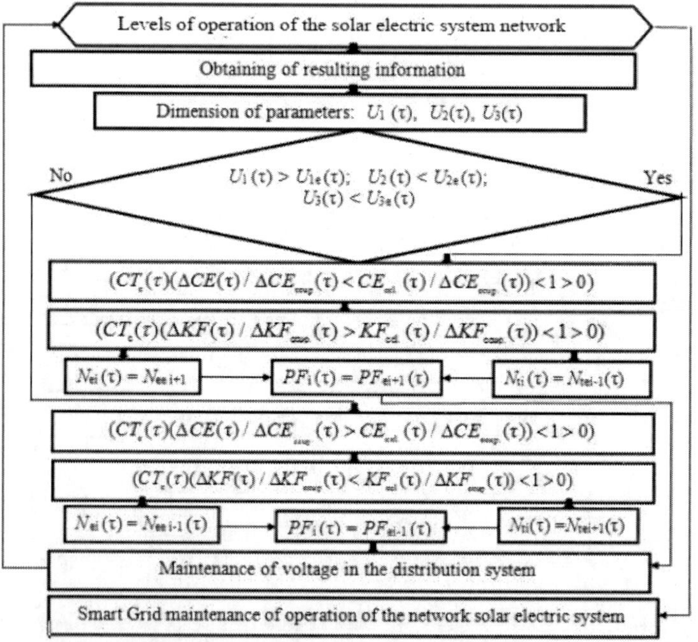

Figure 4.4. Block diagram of maintenance of operation of the solar electric system network: U_1, U_2, U_3 – voltage at the hybrid inverter input and at the frequency converter output, as well as within the distribution system, respectively, volts; *CE* – battery capacity, ampere-hours; *KF* – power factor of the solar electric system network; N_e, N_t – the power of photoelectric module and thermoelectric accumulator, respectively, kW; ι – time, seconds. Indices: *i* – number of operation levels; e – the reference value of the parameter; ccup – the constant calculated value of the parameter of the upper level of operation; ccl – the constant calculated value of the operation level parameter.

Table 4.2. Heat exchange parameters of the battery

Operation levels	Parameter		
	α_1, W/(m^2·K)	α_2,W/(m^2·K)	*k*, W/(m^2·K)
Charge, discharge	8.176	8.152	2.43

Note: α_1– coefficient of heat transfer from the electrolyte to a wall of the battery plate when charged, W/(m^2·K); α_2 – coefficient of heat transfer from the wall of the battery plate to the electrolyte when discharged, W/(m^2·K); k – heat exchange coefficient, W/(m^2·K).

Table 4.3. Time constants and coefficients of mathematical models of dynamics of the solar electric system network

Operation levels	T_e, s	T_m, s	ε	ζ	L, m
Charge	3395	11378	0.997	0.227	1.6
Discharge	3405	11411	0.997	0.227	1.6

Based on the proposed mathematical substantiation for operational maintenance of the Smart Grid solar electric system network (1) to (4) the block diagram for the control of serviceability of the solar electric system network (Figure 4.3) is developed.

Control of solar electric system network operability (Figure 4.3) enables obtaining the resulting information for the benefit of advance decision-making about the maintenance of the voltage in the distribution system. Based on the proposed mathematical substantiation (1) to (4) the block diagram of operational maintenance of the Smart Grid solar electric system network (Figure 4.4) is developed.

Voltage maintenance in the distribution system (Figure 4.4) enables ensuring the operation of the solar electric system network.

Results and Discussion

The Smart Grid System of Operational Maintenance of the Solar Electric System Network at the Decision-Making Level

A comprehensive integrated system has been developed (Table 4.4) for maintaining the operation of the solar electric system network based on a prediction of changes in the battery capacity and power factor of the solar electric system network. Advanced decisions on changing the capacity of the thermoelectric accumulator redistribute the accumulated electrical energy. Changing the level of transmission of electric energy to the network enables maintaining the voltage within the distribution system by maintaining the power factor of the solar electric system network. The voltage at the hybrid inverter input, at the frequency converter output, and within the distribution system takes place is measured continuously. The change in the ratio of voltage at the frequency converter output and voltage in the distribution system is evaluated.

Table 4.4. Integrated system of the solar electric system network

Time, τ, 10^3 s	Change in parameters	$\Delta CE(\tau)$ $/\Delta CE_1(\tau)$	$CE(\tau)$, Ah	$\Delta PF(\tau)$ $/\Delta PF(\tau)_2$	$PF(\tau)$
0	Charge – discharge $U_1 = 160$ V; $U_2 = 230$ V; $U_3 = 400$ V; $m = 0.27$	0.2667	100	0.3382	0.7153
3	Charge – discharge $U_1 = 208$ V; $U_2 = 230$ V; $U_3 = 390$ V; $m = 0.31$	0.3683	97.46	0.3892	0.7357
6	Charge – discharge $U_1 = 232$ V; $U_2 = 230$ V; $U_3 = 385$ V; $m = 0.35$	0.4240	96.07	0.4159	0.7464
9	Charge – discharge $U_1 = 256$ V; $U_2 = 230$ V; $U_3 = 380$ V; $m = 0.39$	0.4835	94.58	0.4418	0.7568
12	Charge – discharge $U_1 = 208$ V; $U_2 = 230$ V; $U_3 = 390$ V; $m = 0.43$	0.3683	97.46	0.3892	0.7357
15	Decision-making $m = 0.47$; $U_1 = 280$ V; $U_2 = 184$ V; $U_3 = 400$ V	0.3673	97.48	0.5255	0.7903
18	Charge – discharge $U_1 = 304$ V; $U_2 = 184$ V; $U_3 = 395$ V; $= 0.51$	0.4082	96.46	0.5432	0.7974
21	Charge – discharge $U_1 = 328$ V; $U_2 = 184$ V; $U_3 = 390$ V; $m = 0.55$	0.4511	95.39	0.5773	0.8110
24	Charge – discharge $U_1 = 352$ V; $U_2 = 184$ V; $U_3 = 385$ V; $m = 0.59$	0.4982	94.22	0.6032	0.8214
27	Charge – discharge $U_1 = 376$ V $U_2 = 184$ V; $U_2 = 380$ V; $m = 0.63$	0.5435	93.09	0.6291	0.8318
30	Decision-making $m = 0.67$; $U_1 = 400$ V; $U_2 = 138$ V; $U_3 = 400$ V	0.4320	95.88	0.7128	0.8653
33	Charge – discharge $U_1 = 424$ V; $U_2 = 138$ V; $U_3 = 395$ V; $m = 0.71$	0.4674	93.23	0.7387	0.8757
36	Charge – discharge $U_1 = 448$ V; $U_2 = 138$ V; $U_2 = 390$ V; $m = 0.75$	0.5037	92.33	0.7646	0.8861
39	Charge – discharge $U_1 = 472$ V; $U_2 = 138$ V; $U_3 = 385$ V; $m = 0.79$	0.5414	91.39	0.7905	0.8965
42	Charge – discharge $U_1 = 496$ V; $U_2 = 138$ V; $U_3 = 380$ V; $m = 0.83$	0.5807	90.41	0.8164	0.9069
45	Decision-making $m = 0.87$; $U_1 = 520$ V; $U_2 = 92$ V; $U_3 = 400$ V	0.4783	92.97	0.9	0.9403
48	Charge – discharge $U_1 = 544$ V; $U_2 = 92$ V; $U_3 = 395$ V; $m = 0,91$	0.5087	92.21	0.9260	0.9507
51	Charge – discharge $U_1 = 568$ V; $U_2 = 92$ V; B$U_3 = 390$ V; $m = 0.95$	0.5400	91.43	0.9519	0.9611
54	Charge – discharge $U_1 = 594$ V; $U_2 = 92$ V; $U_3 = 385$ V; $m = 0,99$	0.5744	90.57	0.9803	0.9725
55	Decision-making $m = 1$; $U_1 = 600$ V; $U_3 = 400$ V	0.4250	94.30	1.0749	1.0103

Note: U1, U2, U3 – voltage at the hybrid inverter input and at the frequency converter output, in the distribution system, respectively, volts; Ne – the power of the solar electric system network, kW; Nt – the power of thermoelectric accumulator, kW; m – the power level of the solar electric system network; CE – battery capacity, ampere-hours; KF – power factor of the solar electric system network; ι – time, seconds. Index: 1 – the constant calculated value of the parameter of the upper level of operation.

The integrated Smart Grid system of maintenance of operation of the solar electric system network (Table 4.4) provides an opportunity to coordinate electric power production and consumption.

Coordination of Electric Power Production and Consumption Based on Voltage Maintenance in the Distribution System

The battery capacity at a specified time point was determined as follows:

$$CE_{i+1}(\tau) = CE_i + \\ + \begin{pmatrix} \Delta CF_{i+1}(\tau) / \Delta CE_1(\tau) - \\ -\Delta CE_i(\tau) / \Delta CE_1(\tau) \end{pmatrix} (CE_2 - CE_1), \quad (18)$$

where CE – the battery capacity, ampere-hours; CE_1, CE_2 – initial and final values of the battery capacity, ampere-hours; τ – time, seconds. Indices: 1 – the constant calculated value of the parameter of the upper level of operation; i – the number of levels of operation of the solar electric system network

The power factor of the solar electric system network at the set time is determined as follows:

$$PF_{i+1}(\tau) = PF_i + \\ + \begin{pmatrix} \Delta PF_{i+1}(\tau) / \Delta PF_1(\tau) - \\ -\Delta PF_i(\tau) / \Delta PF_1(\tau) \end{pmatrix} (PF_2 - PF_1), \quad (19)$$

where PF – the power factor of the solar electric system network; PF_1 and PF_2 – initial and final values of the power factor; τ – time, seconds. Indices: 1– the constant calculated value of the parameter of the upper level of operation; i – the number of levels of operation of the solar electric system network.

For example, in a period of $15 \cdot 10^3$ s (4.17 hrs) the battery capacity was predicted to increase to the level of 97.48 Ah with voltage growth at the hybrid inverter input at the level of 280 V. Value of the battery capacity was determined using the formula (18) as follows (Table 4.4, Figure 4.5):

$$97.48 \text{ Ah} = 94.58 + (0.4835 - 0.3673)(100 - 75).$$

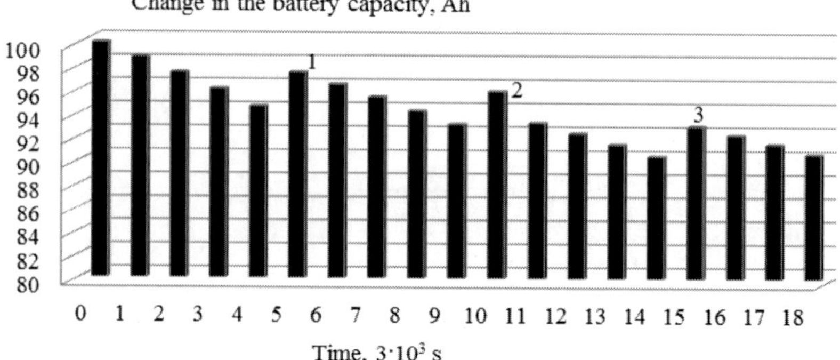

Figure 4.5. Maintenance of change in the battery capacity.
1, 2, and 3 are the points when decisions were made to change the level of power of *the* thermoelectric accumulator.

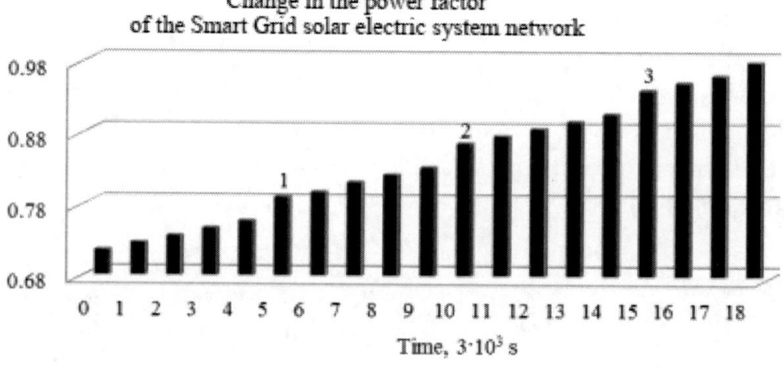

Figure 4.6. Maintenance of change in the power factor. 1, 2, and 3 are the points when decisions were made to change the level of power of the thermoelectric accumulator and transmission of electrical energy to the grid.

In this period of time, it is necessary to make a preliminary decision to reduce the power of the thermoelectric accumulator to the level of 2 kW and increase the level of power of electrical energy transmission to the network from 0.43 to 0.47. The voltage level within the distribution system is set at the level of 400 V and the power factor of the solar electric system grid is at the level of 0.7003.

The value of the power factor in this period of time using formula (19) is determined as follows (Table 4.4, Figure 4.6):

$$0.7903 = 0.7568+(0.5255–0.4418)(0.98–0.58).$$

The value of the power factor in this period was determined as follows using formula (19) (Table 6.4, Figure 6.6):

$$0.7720 = 0.7688+(0.48–0.4720)\ (0.98–0.58).$$

Performing such actions will enable the maintenance of voltage within the distribution system to coordinate the production and consumption of electric power.

Chapter 5

Smart Grid Technology for Maintaining the Operation of a Solar Charging Station

Abstract

The Integrated Smart Grid System of electric power production and consumption harmonization based on a prediction of changes in the battery capacity is developed. The integrated dynamic subsystem of the photoelectric charging station includes the following components: mains, photoelectric solar panels, a hybrid inverter, rechargeable batteries, a two-way Smart Meter and a charger. Advance decisions on the change in power transmission capacity enabled regulating the voltage within the distribution system by maintaining the power factor of the photoelectric charging station. Voltages at the hybrid inverter input and within the distribution system were measured to assess their ratio. Comprehensive mathematical and logical modelling of the photoelectric charging station was carried out based on the mathematical substantiation of architecture and operational maintenance. A dynamic subsystem including such components as mains, a photoelectric module, a hybrid inverter, batteries, a two-way Smart Meter and a charger formed the basis of the proposed technological system. Time constants and coefficients of dynamic mathematical models were determined in terms of the estimation of changes in the battery capacity and power factor of the photoelectric charging station. A functional estimate of changes in the battery capacity and power factor of the photoelectric charging station was obtained. Maintenance of voltage within the distribution system was realized based on the resulting operational data so as to estimate a change in the battery capacity. Advance decision-making has made it possible to raise the power factor of the photoelectric charging station by up to 40% thanks to matching the electric power production with consumption. Operational maintenance of the photoelectric charging station using the developed Smart Grid technology has enabled the prevention of peak loading of the power system due to a 20% reduction in power consumption from the network.

Keywords: photoelectric charging station, rechargeable battery, hybrid inverter, two-way Smart Meter

Introduction

A serious complication occurring in the use of AC charging stations results in a risk of causing peak load in the mains. Under the condition of the growing number of electric vehicles and the irregularity of their charging, there is a need to build charging stations using renewable energy sources. Distributed generation of electric energy using renewable sources requires connection to Smart Grid technologies for integration into the electricity network. Maintaining the power factor of the photoelectric charging station in regard to the redistribution of electricity produced and consumed is an urgent problem for the further development of Smart Grid technologies. To this end, it is necessary to predict changes in the battery capacity and power factor of the photoelectric charging station by measuring the voltages at the hybrid inverter input and within the distribution system to assess their ratio. Advance decisions concerning changes in the level of power transmission to the electricity network enable adjusting the voltage within the distribution system to maintain the balance of active and reactive power without the use of additional devices. The power factor of photoelectric charging stations is maintained by the coordination of electric energy production and consumption. This enables the prevention of peak loads within the electricity network while satisfying the growing consumer demand. For example, the author's work (E. Chaikovskaya, 2017) based on predicting voltage changes in a storage battery is devoted to the connection to Smart Grid technologies. A technology for maintaining a change in the capacity of a storage battery using the measurement of electrolyte temperature in a set of accumulators was presented. The use of an integrated system for estimating a change in voltage based on matching electrochemical and diffusion charging and discharging processes enables making advance decisions to prevent overcharging and unacceptable discharge. The author's work (E. Chaikovskaya, 2019) tackles forecasting changes in the battery capacity for connection to Smart Grid technologies. It presents an integrated system for maintaining the operation of a wind-solar electrical system. Making advance decisions on changing the power of a thermoelectric accumulator is based on establishing a ratio between the voltage measured at the hybrid charge controller input and an inverter output when measuring frequency. A reduction of thermoelectric accumulator charging time by up to 30% was provided based on changing the speed of the circulating pump motor to change the heated water flow rate and temperature.

Optimization of a charging station integrated into a distribution network with the use of renewable energy sources was presented in (Ferro, G., Laureri, F., Minciardi, R., Robba, M., 2019). A target function has been developed. Its minimization is based on the sum of the costs of charging electric vehicles from an external network and the costs associated with service delays. The need to develop infrastructure in connection with establishing the optimal location of charging stations was stated (Jordán, J., Palanca, J., del Val, E., Julian, V., Botti, V., 2021). An agent-oriented approach based on a genetic algorithm was presented. The proposed multi-agent system takes into account the data of activities in social networks and information on mobility in order to establish optimal configurations. A strategy for hybrid power production and charging electric cars was developed (Jiao, Z., Lu, M., Ran, L., Shen, Z.-J. M., 2020) based on an integrated stochastic planning model. Queue optimization and planning for the volume of power generated from non-renewable and renewable energy sources taking into account the irregularity of solar radiation was used. A model of multipurpose optimization of a charging station based on the theory of fuzzy numbers was proposed (Liu, J., Dai, Q., 2020). An algorithm of a swarm of particles was presented to determine the optimal operation of charging stations with the possibility of testing the model in real operating conditions. Optimization of the charging station operation based on mixed integer programming was presented in (Zaher, G. K., Shaaban, M. F., Mokhtar, M., Zeineldin, H. H., 2021). It was solved in the form of a diagram for the day ahead. The purpose of the proposed approach implies maximizing the profit of the charging station owner while satisfying power consumers based on data of charge and discharge and battery replacement during the day. This model takes into account the arrival of customers, changes in the price of electricity from the net, restrictions on connection to the net, and self-destruction of batteries. The work (Elma, O., 2020) addresses the improvement of power consumption management. A technology of fast charge with direct current based on dynamic estimation of a hybrid station was put forward. A decrease in a peak load on the power system during periods of electric vehicle charge and an increase in battery life due to more controlled coordination of the discharge/charge have been established. It was proposed (Dixon, J., Bell, K., 2020) to increase the capacity of batteries and shift the time when charging of electric vehicles takes place to avoid peak loads on the net. It was proposed (Fathabadi, H., 2020) to ensure the stability of the power net with the use of additional accumulation and wind energy. The search for maximum power with a variable step which is applied to both the photoelectric and wind part of the station was used. Moreover, it was proposed to use an auxiliary power source with control of redistribution of produced and consumed energy. The system produces additional electricity when the output of photoelectric and wind energy is less than that required for charging. The system electrolyser produces hydrogen by absorbing additional electrical power available in the system when the production of photoelectric and wind energy exceeds the charging requirement. Thus, an additional energy system acts as a storage tank, adjusting the charge power in accordance with energy consumption. The use of an additional power source and an additional storage capacity relates to the lack of assessment of change in the battery capacity as an integral part of the charging station. Under conditions of distributed electric energy generation, the operation of the battery capacity as an integral part of the circuit design of charging stations becomes fundamental in terms of voltage regulation in the distribution system. The neural model of predicting changes in parameters of the electrical system based on distributed parameters (Jia, Y., Liu, X. J., 2014) does not estimate the change in battery capacity in terms of matching power production and consumption. The presented literature analysis enables the assessment of charging station optimization based on economic and environmental principles of connection to renewable energy sources. Control of electricity consumption and production is carried out with the use of additional devices for voltage regulation in the distribution system, increasing the capacity of batteries, or inclusion of additional storage devices which requires additional costs. The rechargeable battery as a mandatory element of the technological scheme of the photoelectric charging station can become the main element of voltage regulation in the distribution system. This is possible if changes in its capacity are predicted. In this case, the battery becomes the basis for the redistribution of electric energy between the network and the photoelectric module, i.e., it becomes a voltage regulator in the distribution system. Moreover, the assessment of change in the battery capacity makes it possible to maintain the power factor of the photoelectric charging station. Therefore, it was proposed to measure the voltage at the hybrid inverter input and within the distribution system to assess their ratio. Making advance decisions to change the level of power transmission to the network will enable the regulation of voltage within the distribution system to match the production and consumption of energy. The above substantiates the need for further studies in this area.

Methodological and Mathematical Substantiation

Distributed generation of electric energy using renewable sources requires connection to Smart Grid technologies for integration into the electricity network. For example, the author's work (Chaikovskaya, E., 2017) based on predicting voltage changes in a storage battery is devoted to the connection to Smart Grid technologies. A technology for maintaining changes in the capacity of a storage battery using the measurement of electrolyte temperature in a set of accumulators was presented. The use of an integrated system of estimating a change in voltage based on matching electrochemical and diffusion charging and discharging processes enables making advance decisions on boosting to prevent impermissible overcharge and discharge. The author's work (Chaikovskaya, E., 2019) tackles forecasting change in the battery capacity for connection to Smart Grid technologies. It presents an integrated system of maintaining the operation of a wind-solar electrical system. Making advance decisions on changing the power of a thermoelectric accumulator is based on establishing a ratio between the voltage measured at the hybrid charge controller input and an inverter output when measuring frequency. A reduction in the thermoelectric accumulator charge time of up to 30% was provided based on a change in the speed of a circulating pump motor to change the flow rate and temperature of heated water. There are devices for charging electric vehicles which differ from each other by the type of current used and charging time. For example, Mode 3 of a charging station's operation using alternating current makes it possible to charge electric cars of medium power in 4 hours using a 10-kW charger. Fast charge in Mode 4 using direct current restores the capacity of electric car batteries to 80% in half an hour. A serious complication occurring in the use of AC charging stations consists of a risk of peak loading of mains. In conditions of growing number of electric vehicles and irregularity of charge, there is a need to build charging stations using renewable energy sources.

Maintaining the power factor of the photoelectric charging station with regard to redistribution of produced and consumed electricity is an urgent problem of further development of Smart Grid technologies. To this end, it is necessary to predict changes in the battery capacity and power factor of the photoelectric charging station when measuring the input voltage of a hybrid inverter and voltage within the distribution system to assess their ratio. Advance decisions concerning changes in the level of power transmission to the electricity network make it possible to adjust the voltage within the distribution system to maintain the balance of active and reactive power without the use of additional devices. The power factor of photoelectric charging stations is maintained by coordinating the production and consumption of electric energy. This enables the prevention of peak loads in the electricity network under conditions of satisfying the growing consumer demands. One of the main properties of energy systems is the mandatory exchange of substance, energy, and information with the environment.

Thus, the photoelectric charging station is an open integrated dynamic system, the operation of which requires predicting the changes in battery capacity and power factor of the photoelectric charging station when measuring the input voltage of a hybrid inverter and voltage within the distribution system to assess their ratio. The dynamic characteristics of the photoelectric charging station with sufficient accuracy for practice can be described by a finite set of parameters for changes over time, and the spatial coordinate that coincides with the flow direction of the medium, such as changes in battery capacity and power factor of the photoelectric charging station. Therefore, the dynamic description of the photoelectric charging station most fully and multifacetedly characterizes its operation. Thus, it is possible to determine that a real photoelectric charging station is a dynamic system, the mathematical model of which reflects the properties of the transformation of influences, i.e., its dynamic properties. Due to the fact that the photoelectric charging station reflects the dynamic peculiarities due to the nature of reactions to influences, the operational maintenance of the photoelectric charging station should be part of such a technological system, which is based on a dynamic system. Based on the methodological, mathematical, and logical substantiation of the technological systems (Chapter 1) the architecture, mathematical substantiation of the architecture (1), and mathematical substantiation of operational maintenance (2) of the charging station are proposed (Figure 5.1).

A photoelectric charging station is a dynamic system, the operation of which is the reproduction of a change in initial conditions, external and internal influences, and, for example, changes in solar radiation, power consumption for charging electric vehicles, a voltage in the distribution system, etc. Therefore, when designing a photoelectric charging station, an integrated dynamic subsystem is laid down in its base (Figure 5.1). The integrated dynamic subsystem includes the following components: mains, photoelectric solar panels, a hybrid inverter, rechargeable batteries, a two-way Smart Meter and a charger. When representing the system design as the organization of a complex system, it was expanded by building up the dynamic subsystem blocks that forecast the process components around its base. Other components of the technological system include the units for charging, discharging, and functional efficiency estimation in a coordinated interaction with the dynamic subsystem (Figure 5.1).

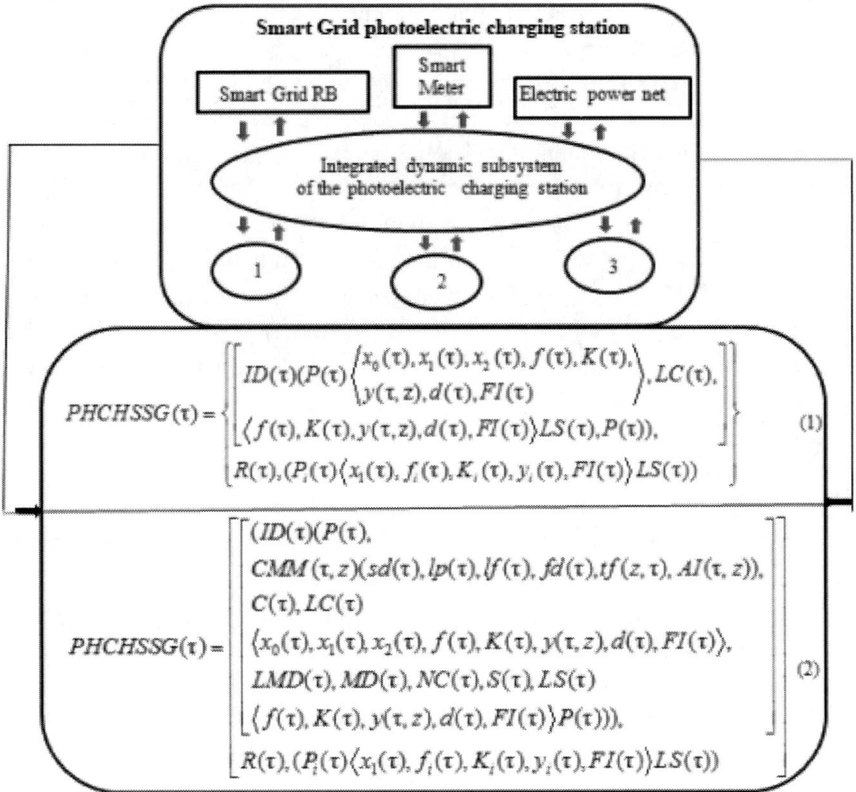

Figure 5.1. Photoelectric charging station Smart Grid: the architecture: RB – rechargeable battery; Smart Meter is a two-way meter detecting changes in the level of power transmission to the network; 1 – the charging unit; 2 – the discharging unit; 3 – the unit for assessing the functional efficiency. Mathematical substantiation of the architecture (1). Mathematical substantiation of operational maintenance (2).

The mathematical substantiation of the architecture of the Smart Grid photoelectric charging station (1), (Figure 5.1), based on the methodology of the mathematical description of power systems dynamics, the method of the graph of cause-effect relations (Chapter 1) is proposed.

Where *PHCHSSG*(τ)– photoelectric charging station Smart Grid; τ – time, seconds; *ID*(τ) – integrated dynamic subsystem (mains, a photoelectric module, a hybrid inverter, a rechargeable battery, a two-way Smart Meter and a charger); $P(\tau)$ – properties of the components of the photoelectric charging station; $x(\tau)$ – impacts (changes in solar radiation, power consumption for charging electric vehicles, voltage in the distribution system, etc; $f(\tau)$ – parameters that are measured: (voltage at the hybrid inverter input, voltage in the distribution system); $K(\tau)$ – coefficients of mathematical description of dynamics of change in the capacity of the storage battery, power factor of the photoelectric charging station; $y(\tau, z)$ – predicted output parameters (battery capacity, power factor of the photoelectric charging station); z – the length coordinate of the battery plates, meters; $d(\tau)$ – dynamic parameters (battery capacity, power factor of the photoelectric charging station); $FI(\tau)$ – the resulting functional information on decision-making; $LC(\tau)$ – logical relations regarding the control of the operability of the photoelectric charging station; $LS(\tau)$ – logical relations regarding the identification of the state of the photoelectric charging station; $R(\tau)$ – logical relations in $PHCHSSG(\tau)$ to confirm the correctness of decisions made from the units of the photoelectric charging station. Indices: i – the number of elements of the photoelectric charging station; 0,1,2 – initial stationary mode, external and internal nature of impacts.

The proposed mathematical substantiation of the operational maintenance of the Smart Grid photoelectric charging station (2), (Figure 5.1) is based on the methodology of the mathematical description of power systems dynamics, and the method of the graph of cause-effect relations (Chapter 1). The basis of the proposed rationale is the mathematical description of the architecture of the Smart Grid photoelectric charging station (1), (Figure 5.1). Prediction of changes in the battery capacity and the power factor of the photoelectric charging station enables making advance decisions to change the level of power transmission to the network in order to maintain voltage in the distribution system. The change in the ratio of voltage at the hybrid inverter input and voltage within the distribution system is assessed. The mathematical substantiation of operational maintenance of the Smart Grid charging station (2) is proposed (Figure 5.1).

Where $PHCHSSG(\tau)$ – operational maintenance of the Smart Grid photoelectric charging station; τ – time, seconds; $ID(\tau)$ – integrated dynamic subsystem (mains, a photoelectric module, a hybrid inverter, a rechargeable battery, a two-way Smart Meter and a charger). $P(\tau)$ – the properties of the elements of the integrated dynamic subsystem, units of the photoelectric charging station; $CMM(\tau,z)$ – complex mathematical modelling of the dynamics of changes in the battery capacity, power factor of the photoelectric charging station; $sd(\tau)$ – the input data (the power of photoelectric module and charge of electric cars, the rechargeable battery and its type and capacity, the two-way Smart Meter and its type and the charger and its type; $lp(\tau)$ – the boundary change in parameters (the voltage at the hybrid inverter input and voltage in the distribution system; $lf(\tau)$ – the levels of operation of the photoelectric charging station; $fd(\tau)$ – the obtained parameters (mode parameters of the photoelectric charging station); $tf(\tau,z)$ – the transfer function of predicted parameters: in the battery capacity, power factor of the photoelectric charging station; $AI(\tau,z)$ – the standard information regarding the evaluation of the maximum admissible change in the battery capacity, power factor of the photoelectric charging station; $C(\tau)$ – the operability control of the photoelectric charging station; $LC(\tau)$ – the logical relations of the operability control of the photoelectric charging station; $x(\tau)$ – impacts (changes in solar radiation, power consumption for charging electric vehicles, voltage within the distribution system, etc.; $f(\tau)$ – the measured parameters: (the voltage at the hybrid inverter input and voltage within the distribution system are measured to assess their ratio); $K(\tau)$ – the coefficients of the mathematical description of the dynamics of a change in the length coordinate of the battery plates; $y(\tau, z)$ – the output parameters (the battery capacity, power factor of the photoelectric charging station); z – the length coordinate of the battery plates, meters; $d(\tau)$ – the dynamic parameters of estimation of a change in the battery capacity, power factor of the photoelectric charging station; $FI(\tau)$ – the functional resulting information on decision-making; $LMD(\tau)$ – the logical relations of decision-making; $MD(\tau)$ – decision-making; $NC(\tau)$ – the new conditions of the photoelectric charging station operation; $S(\tau)$ – the identification of the state of the photoelectric charging station; $LS(\tau)$ – the logical relations of identification of the state of the photoelectric charging station; $R(\tau)$ – the logical relations between the dynamic subsystem and units for charging, discharging, and functional estimation of efficiency that belong to the photoelectric charging station. Indices: i – the number of elements of $PHCHSSG(\tau)$ (τ); 0,1,2 – the initial stationary mode, impacts.

Mathematical substantiation of the architecture of the Smart Grid photoelectric charging station (1) and mathematical substantiation of operational maintenance of the Smart Grid photoelectric charging station (2) (Figure 5.1) enables maintaining the operation of the photoelectric charging station by using the following actions:

- operability control ($C(\tau)$) of the dynamic subsystem based on complex mathematical ($CMM(\tau,z)$) and logical ($LC(\tau)$) modelling regarding obtaining a standard ($AI(\tau,z)$) estimate of a change in the battery capacity, power factor of the photoelectric charging station;
- operability control ($C(\tau)$) of the dynamic system based on complex mathematical ($CMM(\tau,z)$) and logical ($LC(\tau)$) modelling regarding obtaining a functional ($FI(\tau)$) estimate of a change in the battery capacity, power factor of the photoelectric charging station;
- decision-making ($MD(\tau)$) with the use of the resulting functional information ($FI(\tau)$) obtained based on logical modelling ($LMD(\tau)$); decision-making to change the level of power transmission to the network to maintain the power factor of the photoelectric charging station;
- identification ($S(\tau)$) of the new operational conditions of the photoelectric charging station ($NC(\tau)$) based on logical modelling ($LS(\tau)$) as a part of the dynamic subsystem and confirmation of new operating conditions based on logical modelling ($R(\tau)$) from the units of the photoelectric charging station.

Maintaining the Voltage Within the Distribution System Based on a Prediction of Changes in the Battery Capacity

According to formulas (1) and (2), the prediction of a change in the storage battery capacity and power factor of the photoelectric charging station is proposed. The voltage at the hybrid inverter input and voltage within the distribution system are measured to assess their ratio. Transfer functions for "battery capacity – voltage at the hybrid inverter input" and "power factor of the photoelectric charging station – voltage in the distribution system" relations, which were obtained by solving a system of nonlinear differential equations, are presented as follows:

$$W_{CE-U_1} = \frac{K_{ce}K}{(T_e S+1)\beta - 1}\left(1-e^{-\gamma\xi}\right), \tag{3}$$

$$W_{pf-U_2} = \frac{K_{pf}K}{(T_e S+1)\beta - 1}\left(1-e^{-\gamma\xi}\right), \tag{4}$$

where

$$K_{ce} = \frac{I_1 U_1}{(U_1 - U_2)};\ K_{pf} = \frac{I_2 (U_1 - U_2)}{N};\ K = \frac{n(\theta_0 - \sigma_0)}{G_0};$$

$$L = \frac{G_e C_e}{\alpha_0 h_0};\ T_e = \frac{g_e C_e}{\alpha_0 h_0};\ \beta = T_m S + \varepsilon + 1;\ T_m = \frac{g_m C_m}{\alpha_0 h_0};$$

$$\varepsilon = (1 - L);\ \gamma = \frac{(T_e S + 1)\beta\text{-}1}{\beta};\ \xi = \frac{z}{L},$$

where *CE* is the battery capacity, ampere-hours; *PF* – power factor of the photoelectric charging station; I_1, I_2 – currents at the input of the hybrid inverter, in the distribution network, respectively, A; U_1, U_2 – voltages at the hybrid inverter input and in the distribution system, respectively, volts; *N* – the power of the photoelectric charging station, kW; *C* is the specific thermal capacity, kJ/(kg·K); α is the heat transfer factor, kW/(m^2·K); *G* – consumption of the substance, kg/s; *g* – *the* specific weight of a substance, kg/m^3; *h* is the specific surface, m^2/m; σ, θ – the temperature of electrolyte at the battery output and at the distribution wall, respectively, °K; *z* – the length coordinate of the battery plates, meters; $T_{e,}$, T_m – time constants that characterize thermal storage capacity of electrolyte and metal, seconds; *m* is the indicator of the dependence of heat transfer factor on consumption; ι – time, seconds; *S* – the Laplace transform parameter; *S* = ω*j*; ω: frequency, 1/s. Indices: 0 – initial stationary mode; e – electrolyte; m: metal wall.

Transfer functions for "battery capacity – voltage at the input of the hybrid inverter" and "power factor of the photoelectric charging station – voltage in the distribution system" relations were obtained by solving a system of nonlinear differential equations using the Laplace transform. The systems of differential equations include an equation of state as an estimate of the physical model of the photoelectric charging station, an equation of energy of the battery charge and discharge, and a heat balance equation for the battery plate walls. The equation of charge and discharge energies was composed with the representation of a change in electrolyte temperature in the plate pores and above the plates both over time and along the spatial coordinate of the battery plates. The transfer functions include coefficients K_{ce} and K_{pf}, which estimate changes in the battery capacity and the power factor of the photoelectric charging station. When analysing the obtained mathematical model, internal parameters to be diagnosed as being a part of coefficients of the equations of dynamics K_{ce} and K_{pf} were established. In real operating conditions of the power system at a transition from stationary states and under external and internal effects, reorganization of the coefficients of equations of dynamics over time occurs because of a change in the diagnosed internal parameters. The valid parts of the transfer functions are singled out:

$$O(\omega)=\frac{(L_1 A_1)+(M_1 B_1)(1-L)}{(A_1^2+B_1^2)}. \tag{5}$$

The K factor includes the temperature of the separating wall θ:

$$\theta=(\alpha_e(\sigma_1+\sigma_2)/2)+(A(t_1+t_2)/2)/(\alpha_e+A), \tag{6}$$

where σ_1 and σ_2 are electrolyte temperatures at the inlet and outlet of the battery, °K, respectively; t_1 and t_2 are electrolyte temperatures in the plate pores and above the plates at the battery inlet and outlet, respectively, °K; α is the heat transfer factor, kW/(m²·K). Index: e – electrolyte.

$$A=1/(\delta_m/\lambda_m+1/\alpha), \tag{7}$$

where δ is the battery plate wall thickness, meters; λ is the thermal conductivity of the metal of the battery plate, kW/(m·K). Index m is the metal wall of the battery plate.

To use the valid part $O(\omega)$, the following factors were obtained:

$$A_1=\varepsilon-T_e T_m\omega^2; \tag{8}$$

$$A_2=\varepsilon+1; \tag{9}$$

$$B_1=T_e\varepsilon\omega+T_e\omega+T_m\omega; \tag{10}$$

$$B_2=T_m\omega; \tag{11}$$

$$C_1 = \frac{A_1 A_2 + B_1 B_2}{A_2^{\ 2} + B_2^{\ 2}}; \tag{12}$$

$$D_1 = \frac{A_2 B_1 - A_1 B_2}{A_2^{\ 2} + B_2^{\ 2}}; \tag{13}$$

$$L_1 = 1 - e^{-\zeta C_1} \cos\left(-\xi D_1\right); \tag{14}$$

$$M_1 = -e^{-\zeta C_1} \sin\left(-\xi D_1\right). \tag{15}$$

The transfer functions (3) and (4), which were obtained based on the use of the operator method of solving the system of nonlinear differential equations, include the Laplace transform parameter – S (S = ωj), where ω is the frequency, 1/s. To switch from the frequency area to the time area, the valid part (5) was obtained as a result of the mathematical treatment of transfer functions. It is this part that is included in the integrals (16) and (17), which makes it possible to obtain dynamic characteristics of changes in the battery capacity and the power factor of the photoelectric charging station using the inverse Fourier transform:

$$CE(\tau) = \frac{1}{2\pi} \int_0^{\infty} K_{ce} KO(\omega) \sin(\tau\omega/\omega) d\omega, \tag{16}$$

$$PF(\tau) = \frac{1}{2\pi} \int_0^{\infty} K_{pf} KO(\omega) \sin(\tau\omega/\omega) d\omega, \tag{17}$$

where *CE* is the battery capacity, ampere-hours; *PF* is the power factor of the photoelectric charging station.

Thus, for obtaining of reference estimation of change in the battery capacity and power factor of the photoelectric charging station, a block diagram is proposed (Figure 5.2) with, for example, the initial data of a photoelectric charging station with power of 30 kW, for charging electric vehicles with power of 22 kW.

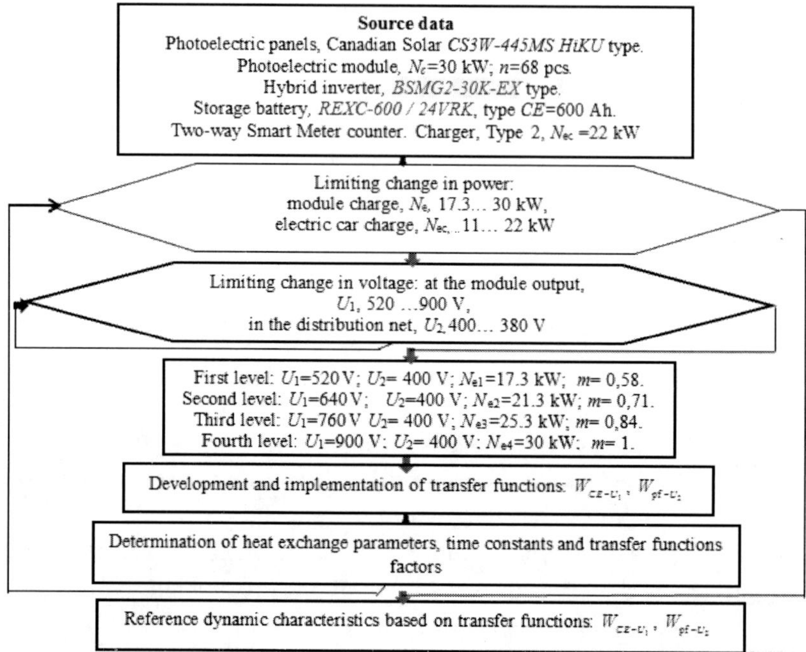

Figure 5.2. Block diagram of comprehensive mathematical modelling of the photoelectric charging station: N_e, N_{ec} – the power of photoelectric module and charge of electric cars, respectively, kW; *CE*– battery capacity, ampere-hours; U_1, U_2 – voltage at the hybrid inverter input and in the distribution system, respectively, volts; *n* – the number of photoelectric panels; *m* – the power level of the photoelectric charging station.

The following levels of operation of the photoelectric charging station have been established for the change in the voltage at the hybrid inverter input and in the distribution system: – first level: 520–400V; second level: 640–400V; third level: 760–400V; fourth level: 900–400V. They correspond to changes in the power of the photoelectric module: 17.3, 21.3, 25.3, 30, respectively, kW and the power level: 0.58, 0.71, 0.84, 1.

According to formulas (1) – (4) and the proposed block diagram (Figure 5.2), the results of reference information obtained on the basis of complex mathematical modelling of the Smart Grid photoelectric charging station are presented (Tables 5.1–5.3).

Table 5.1. Mode parameters of the photoelectric charging station

Levels of operation	N_e, kW	U_1, V	U_2, V	m
first level	17.3	520	400	0.58
second level	21.3	640	400	0.71
third level	25.3	760	400	0.84
Fourth	30	900	400	1

Note: N_e – the power of photoelectric charging station; U_1, U_2 – voltages at the input to the hybrid inverter and in the distribution system, respectively, volts; m – power level of the photoelectric charging station.

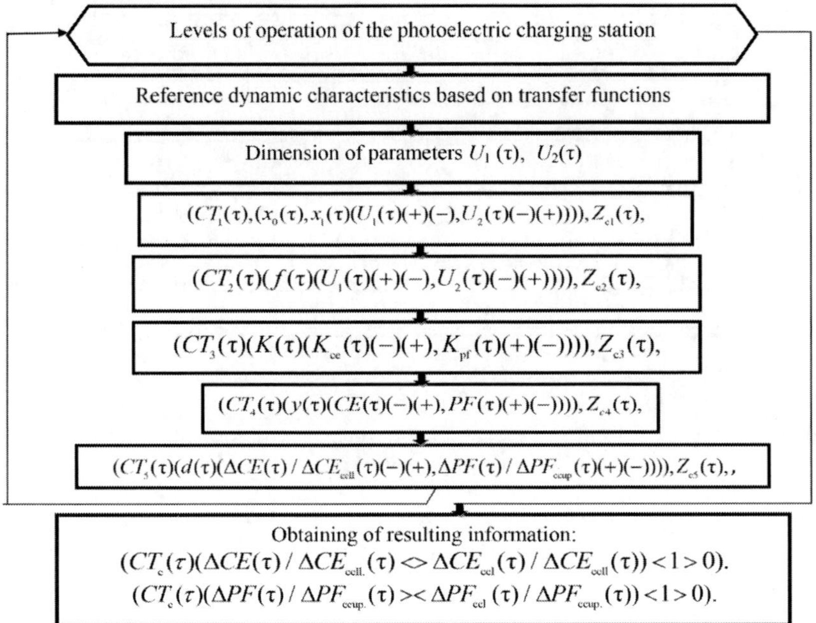

Figure 5.3. Block diagram of the photoelectric charging station operational control: U_1, U_2 – voltages at the input to the hybrid inverter and in the distribution system, respectively, volts; CE – battery capacity, ampere-hours; KF – power factor of the photoelectric charging station; CT – event control; Z – logical relations; d – dynamic parameters; x – effects; f – parameters measured; y – parameters predicted; K – coefficients of mathematical description; ι – time. Indices: c – operability control; ccup, ccll – the constant calculated value of the parameter of upper and lower levels of operation, respectively; ccl – the constant calculated value of the level of operation parameter; 0,1,2 – initial stationary mode, external, internal influences; 3 – coefficients of dynamics equations; 4 – significant predicted parameters; 5 – dynamic parameters.

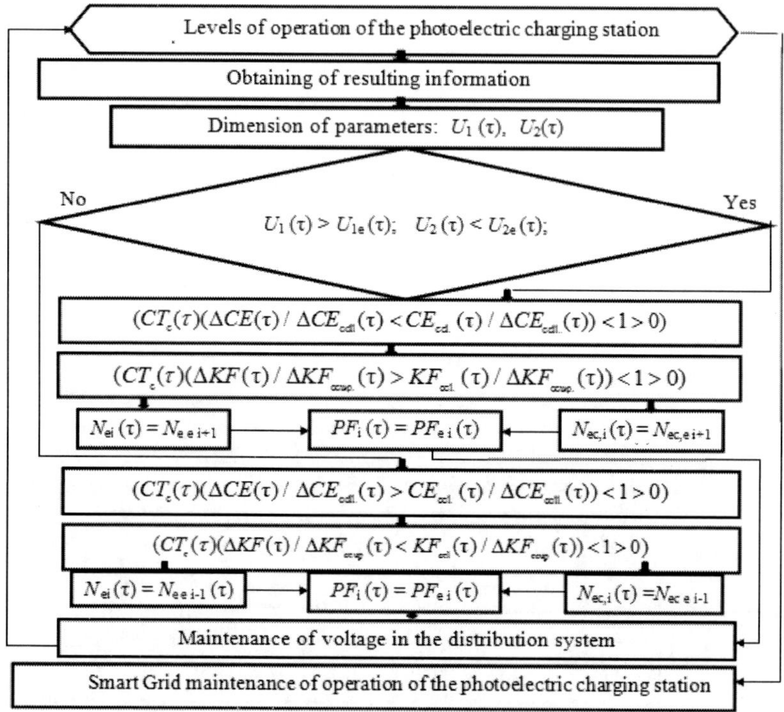

Figure 5.4. Block diagram of operational maintenance of the photoelectric charging station: U_1, U_2 – voltage at the input to the hybrid inverter and in the distribution system, respectively, volts; *CE* – battery capacity, ampere-hours; *KF* – power factor of the photoelectric charging station; N_e, N_{ec} – the power of the photoelectric module and charge of electric cars, respectively, kW; ι – time; Indices: *i* – number of operation levels; r – reference value of the parameter; ccupl, ccll – the constant calculated value of the parameter of the upper and lower levels of operation, respectively; ccl – the constant calculated value of the parameter of the operation level.

Time constants and the coefficients that are components of mathematical models of dynamics (3), (4) presented in Table 5.3 were obtained based on the parameters of heat exchange for charge and discharge of the battery presented in Tables 5.1 and 5.2.

Based on the proposed mathematical substantiation for operational maintenance of the Smart Grid photoelectric charging station (1) to (4) the block diagram for the serviceability control of the photoelectric charging station (Figure 5.3) is developed.

Table 5.2. Heat exchange parameters of the battery

Operation levels	Parameter		
	α_{ch}, W/(m^2·K)	α_{dch}, W/(m^2·K)	*k*, W/(m^2·K)
Charge, discharge	15.298	15.232	3.359

Note: α_{ch} – coefficient of heat transfer from the electrolyte to a wall of the battery plate when charged, W/(m^2·K); α_{dch} – coefficient of heat transfer from the wall of the battery plate to the electrolyte when discharged, W/(m^2·K); k – coefficient of heat exchange, W/(m^2·K).

Table 5.3. Time constants and coefficients of mathematical models of dynamics of the photoelectric charging station

Operation levels	T_e, s	T_m, s	ε	ζ	*L*, m
Charge	1467.56	13352.5	0.973	0.647	35.51
Discharge	1466.94	13346.86	0.973	0.647	35.51

The operability control of the photoelectric charging station (Figure 5.3) enables obtaining the resulting information on decision-making about the maintenance of the voltage in the distribution system.

Based on the proposed mathematical substantiation (1) to (4), the block diagram of operational maintenance of the Smart Grid photoelectric charging station based on maintaining the distribution system voltage was developed (Figure 5.4).

Voltage maintenance within the distribution system (Figure 5.4) makes it possible to ensure the operation of the photoelectric charging station.

Results and Discussion

The Smart Grid System for Maintaining the Operation of the Photoelectric Charging Station at the Decision-Making Level

A comprehensive integrated system has been developed (Table 5.4) for maintaining the operation of the photoelectric charging station based on a prediction of changes in the battery capacity and power factor of the photoelectric charging station. Advance decisions on changes in the level of transmission of electrical energy to the network enable maintaining the voltage within the distribution system by maintaining the power factor of the photoelectric charging station. Continuous measurement of the voltage at the input to the hybrid inverter and within the distribution system takes place to assess their ratio.

Table 5.4. Integrated system for operational maintenance of the charging station

Time, τ, 10^3 s	Change in parameters	$\Delta CE(\tau)$ $/\Delta CE_1(\tau)$	$CE(\tau)$, Ah	$\Delta PF(\tau)$ $/\Delta PF(\tau)_2$	$PF(\tau)$
0	Charge – discharge $U_1 = 520$ V; $U_2 = 400$ V; $N_e = 17.3$ kW; $m = 0.58$	1	600	0.2400	0.6760
3	Charge – discharge $U_1 = 544$ V; $U_2 = 395$ V; $N_e = 18.1$ kW; $m = 0.6$	0.8425	576.38	0.30	0.7000
6	Charge – discharge $U_1 = 568$ V; $U_2 = 390$ V; $N_e = 18.9$ kW; $m = 0.63$	0.7364	560.47	0.3560	0.7224
9	Charge – discharge $U_1 = 592$ V; $U_2 = 385$ V; $N_e = 19.7$ kW; $m = 0.66$	0.66	549.01	0.4140	0.7456
12	Charge – discharge $U_1 = 616$ V; $U_2 = 380$ V; $N_e = 20.5$ kW; $m = 0.68$	0.6023	540.35	0.4720	0.7688
15	Decision-making $m = 0.71$; $U_1 = 640$ V; $U_2 = 400$ V; $N_e = 21.3$ kW	0.6154	542.31	0.48	0.7720
18	Charge – discharge $U_1 = 664$ V; $U_2 = 395$ V; $N_e = 22.1$ kW; $m = 0.74$	0.5696	535.45	0.5380	0.7952
21	Charge – discharge $U_1 = 688$ V; $U_2 = 390$ V; $N_e = 22.9$ kW; $m = 0.76$	0.5328	529.93	0.5960	0.8184
24	Charge – discharge $U_1 = 712$ V; $U_2 = 385$ V; $N_e = 23.7$ kW; $m = 0.79$	0.5025	525.39	0.6540	0.8416
27	Charge – discharge $U_1 = 736$ V; $U_2 = 380$ V; $N_e = 24.5$ kW; $m = 0.82$	0.4771	518.58	0.7120	0.8648
30	Decision-making $m = 0.84$; $U_1 = 760$ V; $U_2 = 400$ V; $N_e = 25.3$ kW	0.4872	520.085	0.72	0.8680
33	Charge – discharge $U_1 = 784$ V; $U_2 = 395$ V; $N_e = 26.1$ kW; $m = 0.87$	0.4450	518.25	0.7779	0.8912
36	Charge – discharge $U_1 = 810$ V; $U_2 = 390$ V; $N_e = 27$ kW; $m = 0.9$	0.4286	516	0.84	0.9160
39	Charge – discharge $U_1 = 834$ V; $U_2 = 385$ V; $N_e = 27.8$ kW; $m = 0.93$	0.4142	513.84	0.8980	0.9392
42	Charge – discharge $U_1 = 858$ V; $U_2 = 380$ V; $N_e = 28.6$ kW; $m = 0.95$	0.4154	514.02	0.9559	0.9624
45	Decision-making $m = 1$; $U_1 = 900$ V; $U_2 = 400$ V; $N_e = 30$ kW	0.4341	511.22	1	0.98

Note: U_1, U_2 – voltage at the input to the hybrid inverter and in the distribution system, respectively, volts; CE – battery capacity, ampere-hours; KF – power factor of the photoelectric charging station; N_e – power of the photoelectric module, kW; m – level of power of the photoelectric charging station; τ – time, Indices: i – number of operation levels; 1,2 – constant calculated value of the parameter of lower and upper levels of operation, respectively.

The integrated system for operational maintenance of the Smart Grid photoelectric charging station (Table 5.4) enables the coordination of electric power production and consumption.

Coordination of Electric Power Production and Consumption Based on Voltage Maintenance in the Distribution System

The battery capacity at a specified point in time was determined as follows:

$$CE_{i+1}(\tau) = CE_i - \\ + \begin{pmatrix} \Delta CF_i(\tau) / \Delta CE_{\text{ccll}}(\tau) - \\ -\Delta CE_{i+1}(\tau) / \Delta CE_{\text{ccll}}(\tau) \end{pmatrix} (CE_2 - CE_1), \quad (18)$$

where CE – the battery capacity, ampere-hours; CE_1, CE_2 – initial and final values of the battery capacity, ampere-hours; τ – time, seconds. Indices: ccll – the constant calculated value of the parameter of lower level of operation; i – the number of levels of operation of the photoelectric charging station.

The power factor of the photoelectric charging station at the set time is determined as follows:

$$PF_{i+1}(\tau) = PF_i + \\ + \begin{pmatrix} \Delta PF_{i+1}(\tau) / \Delta PF_{\text{ccupl}}(\tau) - \\ -\Delta PF_i(\tau) / \Delta PF_{\text{ccupl}}(\tau) \end{pmatrix} (PF_2 - PF_1), \quad (19)$$

where PF – power factor of the photoelectric charging station; PF_1, PF_2 – initial and final values of the power factor; τ – time, s. Indices: ccupl – constant calculated value of the parameter of the upper level of operation; i – the number of levels of operation of the photoelectric charging station.

For example, in a period of $15 \cdot 10^3$ s (4.17 hrs), the battery capacity was predicted to increase to the level of 542.31 Ah with voltage growth at the input to the hybrid inverter at the level of 540 V. The value of the battery capacity was determined using the formula (18) as follows (Table 6.4, Figure 6.5):

542.31 Ah = 540.35+(0.6154–0.6023) (600–450).

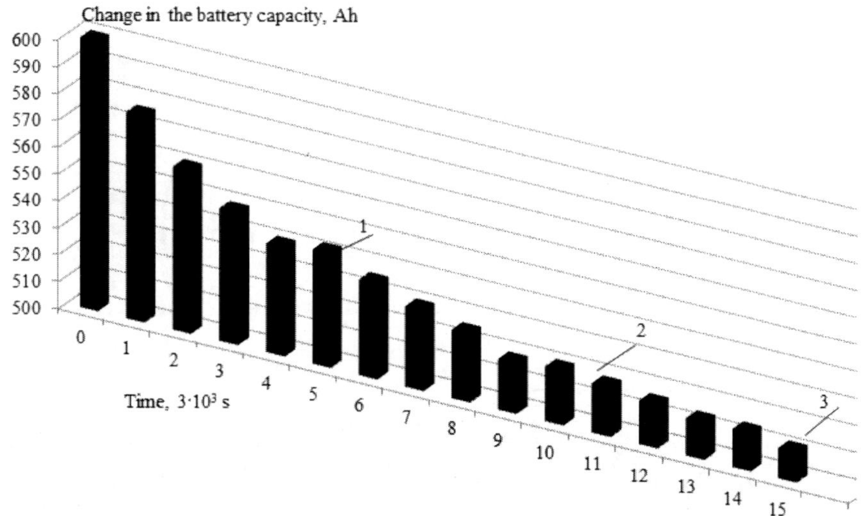

Figure 5.5. Maintenance of change in the battery capacity. 1, 2, and 3 are the points when decisions were made to change the level of power transmission to the network.

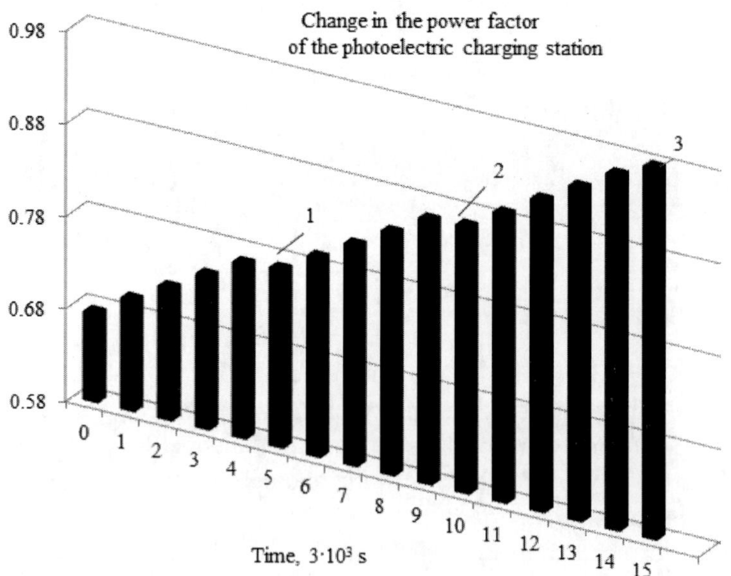

Figure 5.6. Maintenance of operation of the photoelectric charging station. 1, 2, and 3 are the points when decisions were made on the change in the level of power transmitted to the network.

During this period, it was necessary to make an advance decision to raise the level of power transmission to the network from 0.68 to 0.71. The voltage level within the distribution system was set at 400 V and the power factor of the photoelectric charging station was at 0.7720.

The value of the power factor in this period was determined using formula (19) (Table 6.4, Figure 6.6) as follows:

$$0.7720 = 0.7688+(0.48-0.4720)\ (0.98-0.58).$$

Performing such actions will enable the maintenance of voltage within the distribution system to coordinate the production and consumption of electric power.

Chapter 6

Smart Grid Technology for Maintaining the Operation of Autonomous Wind-Solar Electric System

Abstract

The developed integrated Smart Grid system for supporting the operation of a wind-solar electric system is based on predicting a change in the capacity of a rechargeable battery by measuring the voltage at the hybrid charge controller input, the voltage at the inverter output, and the current frequency. Making advance decisions to support the capacity of a rechargeable battery related to a change in the capacity of a thermoelectric battery is based on establishing the ratio of voltage measured at the hybrid charge controller input to voltage at the inverter output. A change in the rotational speed of the electric motor of the circulating pump has been ensured in terms of changes in consumption and the temperature of heated water by reducing charge duration by up to 30%. The basis for the proposed technological system is a dynamic subsystem that includes the following components: a wind-energy installation, a photoelectrical electrical module, a hybrid charge controller, an inverter, an array of rechargeable batteries, and a thermoelectric battery. A functional assessment has been derived for a change in the capacity of a rechargeable battery, the rotational speed of the electric motor of the circulating pump, and local water consumption related to a change in the local water temperature in the range of 30–70°C. Defining the resulting functional information on forecasting a change in the capacity of a rechargeable battery enables making the following advance decisions on changing the rotational speed of the electric motor of the circulating pump and the consumption of local water. Maintaining the capacity of a rechargeable battery is carried out based on adjusting the generation and consumption of energy.

Keywords: wind-solar electrical system, rechargeable battery, thermoelectric battery, hybrid charge controller, inverter

Introduction

Given the need to save natural fuel and reduce harmful emissions into the atmosphere, supporting the operation of wind-solar electrical systems requires improvements in terms of accumulating electrical energy. Thus, for example, based on the mathematical and logical modelling related to the technological system for operating a rechargeable battery, the technology has been devised to support a change in the battery capacity, based on predicting a voltage change when measuring the temperature of the electrolyte in the volume of rechargeable batteries. Application of the integrated evaluation system of voltage change, obtained on the basis of adjusting the electrochemical and diffusion processes of discharge and charge, enables making timely decisions about additional charging to prevent recharge and inadmissible discharge (Chapter 2).

Owing to the combination of two different energy sources into a single wind-solar electric system, predetermined by the need to expand the duration of electric energy consumption, specialized hybrid controllers have been designed to control the charging of rechargeable batteries. PWM controllers' operation is based on the pulse-width modulation of the charge current. MPPT controllers that operate on the principle of finding the maximal power of energy generation adjust a change in voltage to the change in current. Known methods for charging MPPT controllers are based on scanning aimed at establishing a point of maximum power. Thus, according to the perturb and observe method, the controller performs a full scan of the volt-ampere characteristic, thereby finding the point of maximum power over a specified time duration. Until the next full scanning, the controller calculates a change in power and sets the operating point to a new voltage if the power at it is higher. Improvement of the method for determining the optimal maximum power is performed at the level of assessing a change in the cycle of complete scanning, depth, and frequency of iterations, etc. According to the scan and hold method, following the primary scanning, the defined maximum power point does not change until the next full scanning. The method of percentage of open circuit voltage enables using the operating voltage based on the estimation of idling voltage. The choice of a working point without scanning is also supported by the method to evaluate the maximum power of energy generation in compliance with the maximum capacity of a rechargeable battery. Under conditions of unstable solar radiation and changing wind speed, obtaining the total maximum power of two energy sources does not coincide in time, neither over the day or over the year. Under difficult conditions for electrical energy generation at changing consumption, it is a relevant task, related to the further development of the wind-solar electrical system technologies to support the capacity of a rechargeable battery in terms of adjusting the generation and consumption of energy to maintain energy efficiency. To this end, it is necessary to predict the change in the capacity of a rechargeable battery by measuring the voltage at the hybrid charge controller input, as well as the voltage at the inverter output to estimate their ratio. Making advance decisions about a change in the power of a thermoelectric battery makes it possible, while maintaining the capacity of the rechargeable battery, to change the rotational speed of the electric motor of the circulating pump to change local water consumption and the heating level. Known methods for the optimization of wind-solar electrical systems are based on improving intelligent management systems based on hybrid charge controllers, emphasizing the relevance of combining two different sources of energy. One direction to optimize wind-solar electrical systems is to improve the function of MPPT controllers in terms of finding maximum power. For example, a paper (Mujahed, A., Nassef, A., & Rezk, H., Nisar, K., 2018) proposed an MPPT algorithm for hybrid controllers based on integration between management of fractional order and a method of incremental conductivity to improve the accuracy of monitoring. A study (Shahriari, M., Blumsack, S., 2018) established the possibility of obtaining maximum power in a different period, both during the day and over the year; an article (Fathabadi, H., 2017) underlined the maximum efficiency of hybrid controllers MPPT in terms of defining maximum power at efficiency of up to 99.60%. Paper (Chandra, M., Naiki, S., & Anush, D., 2017) has established, based on the proposed technological scheme, the increase in the efficiency of a wind-solar electric system in comparison with a separate application of a wind-solar electric system employing the MPPT function of hybrid controllers. Authors (Cao, Z., O'Rourke, F., & Lyons, W., 2017) suggested a specialized algorithm that was developed based on the simulation of subsystems of wind and solar power, rechargeable battery, and inverter, based on MATLAB Simulink, to improve the MPPT function of hybrid controllers. A paper (Vasant, L., Pawar, V., 2017) proposed using a constant voltage method to maximally transmit power based on the MPPT function of hybrid controllers. The conclusion was drawn about the necessity to define key features in the application of the method. Improvement of intelligent control systems has been addressed in work (Derrouazin, A., Aillerie, M., & Mekkakia-Maaza, N., Charles, J., 2017), which presented a hybrid controller with fuzzy logic to manage the generation and transmission of energy. The authors described command signals, as well as determined the ratio of renewable energy system integration. However, the proposed technology does not include aligning the generation and consumption of electrical energy, which is why the excess energy was directed into the system of electrolysis to obtain hydrogen. More to the point, the reported technology requires a network connection to ensure peak consumption. A paper (Yubin, J., Liu, X.,2014) suggested using a neural model to forecast changes in the parameters for a wind-solar electric system as a nonlinear dynamic system. The optimization of predictive control is based on the representation of the subsystems that generate wind energy, solar energy, as well as rechargeable battery and load. Based on computer simulation, the authors established the performance efficiency of the intelligent control over a distributed model but failed to assess a change in the capacity of a rechargeable battery in terms of adjusting the generation and consumption of energy. A study (Saidi, A., Chellali, B., 2017) reported the optimization of a wind-solar electric system based on control over inverter operation. The authors introduced an additional module of dispatching control to track the maximum power of energy sources. They used special methods for the simultaneous collection of energy under different climatic conditions. For a photovoltaic system, the perturb and observe algorithm; for a wind turbine, Hill Climb Search. Maintaining the consumption of generated energy based on establishing constant voltage and current frequency does not estimate a change in the capacity of a rechargeable battery. A paper (Maleki, A., Hafeznia. H., & Rosen, M., Pourfayaz, F., 2017) proposed an economic model for the optimization of a wind-solar electric system based on minimizing operating costs. The authors applied the methods of meta-heuristic optimization to compare a genetic algorithm and the Particle Swarm Optimization algorithm. The optimization takes into consideration the costs of production of solar and wind energy, recovery of thermal energy from a fuel cell, energy supply, electricity exchange in a network, as well as maintenance. The authors established the economic feasibility of the structure of a hybrid combined wind-solar electric system, connected to the grid and with a heat element, but at the static level. They established the benefits of using a genetic algorithm for obtaining the optimization results, but without predicting the change of parameters for the system's components. Based on analysis of the scientific literature it was found that the hybrid charge controller as part of a wind-solar electric system maintains the charge of a rechargeable battery by using a thermoelectric battery as a non-regulated ballast. Resetting the excess energy to the ballast when using the MPPT functions of the controller leads to unrecoverable losses of electrical energy, which does not make it possible to ensure an appropriate level of the charge capacity of a rechargeable battery. Moreover, the use of a thermoelectric battery as a ballast eliminates the need to maintain the operation of a wind energy installation by using the accumulation of heat to regulate the power of a wind turbine. Failure to consider this property of a thermoelectric battery could lead to the acceleration of a wind turbine at considerable wind speed and to its malfunctioning. It is the thermoelectric battery that appears to be the main centre of adjusting a change in the total power of a wind-solar electric system to power consumption through redistribution of the accumulated heat and electric energy in terms of consumption. Therefore, it is proposed to measure the total voltage at the hybrid charge controller input and voltage at the inverter output to estimate the ratio of electrical energy generation to its consumption when measuring the current frequency. It is known that a thermoelectric battery is executed in line with a thermostat principle, that is, when determining the required temperature for heating local water, the thermoelectric battery is disconnected from power. Not using a change in the local water consumption during the charging of the thermoelectric battery increases the charging duration and leads to considerable costs related to electricity consumption. Therefore, to maintain the operation of the wind-solar electrical system, it is necessary to predict the change in the capacity of the battery. In order to save energy, it is necessary to measure the voltage at the hybrid charge controller input and voltage at the inverter output to estimate their ratio when measuring the current frequency. Making advance decisions about changing the power of a thermoelectric battery makes it possible, while maintaining the capacity of the rechargeable battery, to change the rotational speed of the electric motor of the circulating pump. Changing the flow rate of the local water supplied to a thermoelectric battery ensures the set level of change in the temperature of heated local water.

Methodological and Mathematical Substantiation

Distributed generation of electric energy using renewable sources requires connection to Smart Grid technologies for integration into the electric network. For example, the author's work (Chapter 2) based on predicting voltage changes in a storage battery is devoted to connection to Smart Grid technologies. A technology for maintaining change in the capacity of a storage battery during the measurement of electrolyte temperature in a set of accumulators was presented. The use of an integrated system of estimating a change in voltage based on matching electrochemical and diffusion processes of charge and discharge enables making advanced decisions on boosting to prevent impermissible overcharge and discharge. One of the main properties of energy systems is the mandatory exchange of substance, energy and information with the environment. Thus, the wind-solar electric system – open integrated dynamic system, the operation of which requires predicting changes in battery capacity when measuring voltage at the hybrid charge controller input, voltage at the inverter output and the current frequency. The ratio of voltages at the hybrid charge controller input and the inverter output is measured. The dynamic characteristics of the wind-solar electric system with sufficient accuracy for practice can be described by a finite set of parameters for changes in time, and the spatial coordinate that coincides with the direction of flow of the medium, such as changes in battery capacity and the capacity of a thermoelectric battery. Therefore, the dynamic description of the wind-solar electric system most fully and multifacetedly characterizes its operation. Thus, it is possible to determine that the real autonomous wind-solar electric system is a dynamic system, the mathematical model of which reflects the properties of the transformation of influences, i.e., its dynamic properties (Chapter 1). Because the wind-solar electric system reflects the dynamic peculiarities due to the nature of reactions to influences, the supporting of the operation of the wind-solar electric system should be part of such a technological system, which is based on a dynamic system. Based on the methodological, mathematical, and logical substantiation of the technological systems architecture (Chapter 1), mathematical substantiation of the architecture (1) and operational maintenance (2) of the Smart Grid wind-solar electric system is proposed (Figure 6.1).

A wind-solar electric system is a dynamic system, the operation of which is the reproduction of a change in external, internal influences, and initial conditions, for example, changes in solar radiation, wind speed, electrical energy consumption, etc. Therefore, when designing a wind-solar electric system, an integrated dynamic subsystem is laid down in its base (Figure 6.1). The integrated dynamic subsystem includes the following components: a wind-energy installation, a photoelectric module, a hybrid charge controller and an inverter, an array of rechargeable batteries, and a thermoelectric battery. When representing the system design as the organization of a complex system, it was expanded by building up the dynamic subsystem blocks that forecast the process components around its base. Other components of the technological system include the units of charge and discharge and functional efficiency estimation in a coordinated interaction with the dynamic subsystem (Figure 6.1).

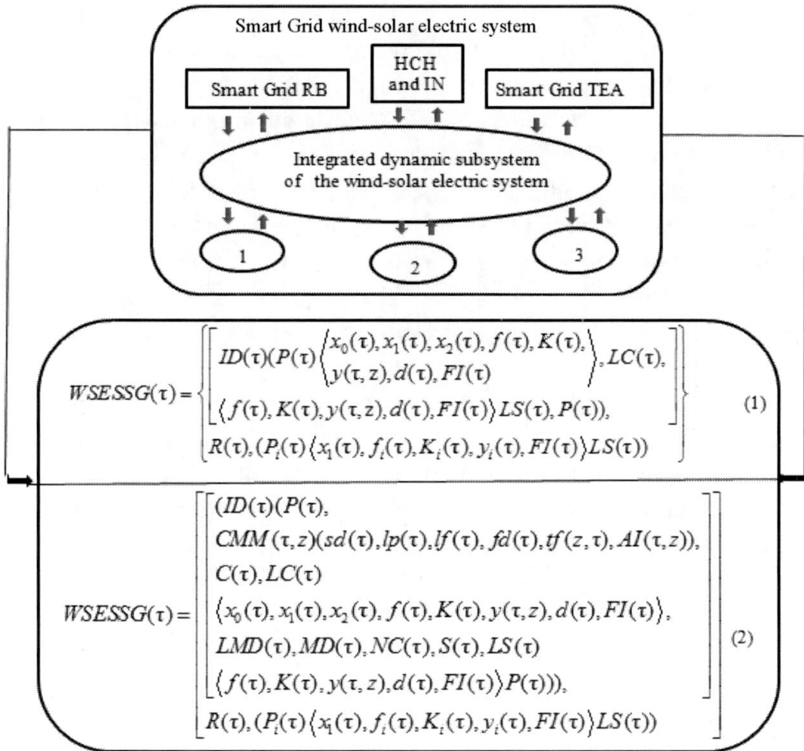

Figure 6.1. Smart Grid wind-solar electric system: the architecture: RB – rechargeable battery; HCH and IN – hybrid charge controller and inverter; TEA – thermoelectric accumulator; 1 – the charging unit; 2 – the discharging unit; 3 – the unit of assessing the functional efficiency. Mathematical substantiation of the architecture (1). Mathematical substantiation for operational maintenance (2).

The mathematical substantiation of the architecture of the Smart Grid wind-solar electric system (1), (Figure 6.1), based on the methodology of the mathematical description of dynamics of power systems, the method of the graph of cause-effect relations (Chapter 1) is proposed.

Where *WSESSG*(τ) – Smart Grid wind-solar electric system; τ – time, seconds; *ID*(τ) – integrated dynamic subsystem (a wind-energy installation, a photoelectric module, a hybrid charge controller and an inverter, an array of rechargeable batteries, a thermoelectric battery); $P(\tau)$ – properties of the components of the wind-solar electric system; $x(\tau)$ – impacts (changes in solar radiation, wind speed, electrical energy consumption, etc; $f(\tau)$ – parameters that are measured: (voltage at the hybrid charge controller input, voltage at the inverter output, current frequency); $K(\tau)$ – coefficients of mathematical description of dynamics of change in the capacity of a rechargeable battery, the rotational speed of the electric motor of the circulating pump, changes in the local water consumption; $y(\tau,z)$ – predicted output parameters (change in capacity of a rechargeable battery, rotational speed of the electric motor of the circulating pump, local water consumption) z – is the coordinate of the heater's length, metres; $d(\tau)$ – dynamic parameters for change in the capacity of a rechargeable battery; the rotational speed of the electric motor of the circulating pump, local water consumption); $FI(\tau)$ – functional resulting information on decision-making; $LC(\tau)$ – logical relations regarding the control of the wind-solar electric system operability; $LS(\tau)$ – logical relations regarding the identification of the state of the wind-solar electric system; $R(\tau)$ – logical relations in $AWSESSG(\tau)$ to confirm the correctness of decisions made from the units of the Wind-Solar electric system. Indices: i – the number of elements of the wind-solar electric system; 0,1,2 – initial stationary mode, external, internal nature of impacts.

The mathematical substantiation for operational maintenance of the Smart Grid wind-solar electric system (2), (Figure 6.1), based on the methodology of the mathematical description of dynamics of power systems, the method of the graph of cause-effect relations (Chapter 1) is proposed. The basis of the proposed rationale is the mathematical description of the architecture of the wind-solar electric system Smart (1), (Figure 6.1). Prediction of changes in the capacity of a rechargeable battery, the rotational speed of the electric motor of the circulating pump, and local water consumption enables making advanced decisions to change the capacity of a thermoelectric battery. A change in the rotational speed of the electric motor of the circulating pump has been ensured in terms of changes in consumption and the temperature of heated water. The mathematical substantiation for operational maintenance of the Smart Grid wind-solar electric system (2) is proposed (Figure 6.1).

Where $WSESSG(\tau)$ – operational maintenance of the Smart Grid wind-solar electric system; τ – time, seconds; $ID(\tau)$ – integrated dynamic subsystem (a wind-energy installation, a photoelectric module, a hybrid charge controller and an inverter, an array of rechargeable batteries, a thermoelectric battery). *P*(τ) – the properties of the elements of the integrated dynamic subsystem, units of the wind-solar electric system; *CMM*(τ,z) – complex mathematical modelling of the dynamics of changes in the battery capacity, the rotational speed of the electric motor of the circulating pump, changes in the local water consumption of the wind-solar electric system; *sd*(τ) – the input data (the power of a wind-energy installation; the power of a photoelectric module; the type of a hybrid charge controller and an inverter, the capacity of a rechargeable battery; the power and volume of a thermoelectric battery; *lp*(τ) – limit to change in parameters (voltage at the hybrid charge controller input, thermoelectric battery's power); *lf*(τ) – levels of operation according to a change in voltage at the hybrid charge controller input and at the inverter output; *fd*(τ) – the obtained parameters (parameters for heat transfer in a thermoelectric battery, the rotational speed of the electric motor of the circulating pump, the flow rate of heated water, the charge duration of a thermoelectric battery for the established levels of operation; *AI*(τ,*z*) – the standard information regarding the evaluation of the maximum admissible change in the battery capacity, rotational speed of the electric motor of the circulating pump, local water consumption; *C*(τ) – the operability control of the wind-solar electric system; *LC*(τ) – the logical relations of the control of the wind-solar electric system operability; *x*(τ) – impacts (change in solar radiation, change in wind speed, change in electricity consumption, etc.; *f*(τ) – the measured parameters: (voltage at the hybrid charge controller input, voltage at the inverter output, current frequency); *K*(τ) – the coefficients of the mathematical description of the dynamics of change in the capacity of a rechargeable battery, the rotational speed of the electric motor of the circulating pump, changes in the local water consumption; *y*(τ,z) – the output parameters (change in capacity of a rechargeable battery, rotational speed of the electric motor of the circulating pump, local water consumption); *z* – is the coordinate of the heater's length, metres; *d*(τ) – the dynamic parameters of estimation of a change in the battery capacity, the rotational speed of the electric motor of the circulating pump, changes in the local water consumption; *FI*(τ) – functional resulting information on decision-making; *LMD*(τ) – the logical relations of decision-making; *MD*(τ) – decision-making; *NC*(τ) – the new conditions of the wind-solar electric system operation; *S*(τ) – the identification of the state of the wind-solar electric system; *LS*(τ) – the logical relations of identification of the state of the wind-solar electric system; *R*(τ) – the logical relations between the dynamic subsystem and units of charge, discharge, functional estimation of efficiency that belong to the wind-solar electric system. Indices: i – the number of elements of $WSESSG(\tau)$; 0,1,2 – the initial, external, and internal character of influences.

Mathematical substantiation of the architecture of the Smart Grid wind-solar electric system (1) and mathematical substantiation for operational maintenance of the Smart Grid wind-solar electric system (2) (Figure 6.1) enable maintaining the operation of the wind-solar electric system using the following actions:

- operability control ($C(\tau)$) of the dynamic subsystem based on complex mathematical ($CMM(\tau,z)$) and logical ($LC(\tau)$) modelling regarding obtaining standard ($AI(\tau,z)$) estimate of a change in the battery capacity of the wind-solar electric system;
- operability control ($C(\tau)$) of the dynamic system based on complex mathematical ($CMM(\tau,z)$) and logical ($LC(\tau)$) modelling regarding the obtaining functional ($FI(\tau)$) estimate of a change in the battery capacity of the wind-solar electric system;
- decision-making ($MD(\tau)$) with the use of the functional resulting information ($FI(\tau)$), obtained based on logical modelling ($LMD(\tau)$); decision-making on changing the current frequency using the functional estimate of the change in the capacity of a rechargeable battery, the rotational speed of the electric motor of the circulating pump, local water consumption;
- identification ($S(\tau)$) of the new operating conditions of the wind-solar electric system ($NC(\tau)$) based on logical modelling ($LS(\tau)$) as a part of the dynamic subsystem and confirmation of new operating conditions based on logical modelling ($R(\tau)$) from the units of the wind-solar electric system.

Mathematical Modelling of the Dynamics of Change in the Capacity of a Rechargeable Battery

According to formulas (1), (2), and mathematical substantiation of the architecture of a technological system (Figure 6.1), it has been proposed to predict a change in the capacity of a rechargeable battery when measuring voltage at the hybrid charge controller input, when measuring voltage at the inverter output and the current frequency. A transfer function for the "capacity of rechargeable battery – power of thermoelectric battery" relation was derived from solving a system of nonlinear differential equations. A change in the capacity of a rechargeable battery when using the estimate of the change in the local water temperature both in time and along the coordinate of the length of the heater in a thermoelectric battery, takes the following form:

$$W_{ec-Nm_1} = \frac{K_{ec}K_w\left(1-L_w^*\right)}{(T_wS+1)\beta-1}\left(1-e^{-\gamma\xi}\right), \tag{3}$$

where

$$K_{ec} = \frac{IU_1}{(U_1-U_2)};\; K_w = \frac{m\left(\theta_0-t_0\right)}{G_{w0}};\; L_w^* = \frac{1}{L_w+1};\quad L_w = \frac{G_wC_w}{\alpha_{w0}h_{w0}};\quad \varepsilon^* = (1-L_w^*);$$

$$\gamma = \frac{(T_wS+1)\beta-1}{\beta};\; \xi = \frac{z}{L_w};\; T_w = \frac{g_wC_w}{\alpha_{w0}h_{w0}};\; \beta = T_mS+\varepsilon^*+1;\; T_m = \frac{g_mC_m}{\alpha_{w0}h_{w0}},$$

where EC is the capacity of a rechargeable battery, ampere-hours; N is the power of a thermoelectric battery, kW; I is current, A; U_1, and U_2 are the voltages at the hybrid charge controller input and the inverter output, respectively, volts; C is the specific heat capacity, KJ/(kg·K); α is the coefficient of heat transfer, kW/(m^2·K); G is the substance flow rate, kg/s; g is the specific mass of a substance, kg/m; h is the specific surface, m^2/m; t, θ are the temperatures of local water, separating wall, respectively, K; z is the coordinate of the heater's length, metres; T_w, T_m are the time constants, characterizing the thermal accumulating capacity of local water, metal, seconds; m is the indicator of dependence of heat transfer coefficient on flow rate; ι is time, seconds; S is the parameter of Laplace transformation; $S = \omega j$; ω is frequency, 1/s. Indices: 0 – original stationary mode; 1 – input to an electric system; w – local water; m – metallic wall.

A transfer function for the "capacity of rechargeable battery – power of thermoelectric battery" relation was derived from solving a system of nonlinear differential equations using a Laplace transform. The system of differential equations includes an equation of state as an estimate of the physical model of an electrical system. The system of differential equations also includes equations of energy in the transmitting and receiving environments – the heater of a thermoelectric battery and local water, respectively, as well as the equation of thermal balance for the wall of a thermoelectric heater.

The equation of energy for the receiving environment was built to represent a change in the local water's temperature both over time and the spatial length coordinate, which coincides with the direction of the environmental flow and includes coefficient K_w. The equation of energy for the transmitting environment includes the coefficient K_{ec}, which assesses a change in the capacity of a rechargeable battery to estimate the ratio of energy generation and consumption.

Mathematical Modelling of the Dynamics of Change in the Rotational Speed of the Electric Motor of the Circulating Pump

According to formulas (1), (2), and the mathematical substantiation of the architecture of a technological system (Figure 6.1), it has been proposed to estimate a change in the rotational speed of the electric motor of the circulating pump in a thermoelectric battery when measuring the current frequency. A transfer function for the "rotational speed of the electric motor of the circulating pump – current frequency" relation takes the following form:

$$W_{n-f_1} = \frac{K_f \chi_t S}{\gamma}\left(1 - e^{-\gamma_1 \xi}\right), \tag{4}$$

where

$$K_f = \frac{120 f(1-s)}{p_n}; \quad \chi_t = -f_s \frac{\partial \rho}{\partial t}; \quad \gamma = \frac{(T_w S + 1)\beta - 1}{L_w \beta}; \; T_w = \frac{g_w C_w}{\alpha_{w0} h_{w0}};$$

$$L_w = \frac{G_w C_w}{\alpha_{w0} h_{w0}}; \; \beta = T_m S + \varepsilon^* + 1; \; T_m = \frac{g_m C_m}{\alpha_{w0} h_{w0}}; \; \varepsilon^* = (1 - L_w^*); \quad L_w^* = \frac{1}{L_w + 1};$$

$$\gamma_1 = \frac{(T_w S + 1)\beta - 1}{\beta}; \; \xi = \frac{z}{L_w},$$

where n is the rotational speed of the electric motor of the circulating pump, rpm; f is the current frequency, Hz; p_n is the number of pairs of poles in the electric motor of the circulating pump; α is the coefficient of heat transfer, kW/(m^2·K); t is the temperature of local water, K; ρ is the density of local water, kg/m^3; *fs* is the bypass for local water, m^2; C is the specific heat capacity, KJ/(kg·K); G is the consumption of a substance, kg/s; ρ is the density of local water, kg/m^3; g is the specific mass of a substance, kg/m; h is the specific surface, m^2/m; z is the coordinate of length of the heater, metres; T_w, T_m are the time constants, characterizing the thermal accumulating capacity of local water, metal, seconds; S is the parameter of Laplace transformation; $S = \omega j$; ω is frequency, 1/s. Indices: 0 – original stationary mode; 1 – input to an electric system; w – local water; m – metallic wall.

Mathematical Modelling of the Dynamics of Change in Local Water Consumption

According to formulas (1), (2), and mathematical substantiation of the architecture of a technological system (Figure 6.1), it has been proposed to estimate a change in the local water consumption when changing the rotational speed of the electric motor of the circulating pump. To this end, a system of differential equations, which includes a local energy equation and an equation of thermal balance for the wall of a heater, is supplemented with the local water equation of continuity. The result of solving a system of differential equations using a Laplace transform is the transfer function for the "local water flow rate – rotational speed of the electric motor of the circulating pump" relation, which estimates a change in water flow rate when changing power of a thermoelectric battery:

$$W_{Gw-n_1} = \frac{\chi_t S}{\gamma}\left(1 - e^{-\gamma_1 \xi}\right), \tag{5}$$

where

$$\chi_t = -f_s \frac{\partial \rho}{\partial t};\ \gamma = \frac{(T_w S + 1)\beta - 1}{L_w \beta};\ T_w = \frac{g_w C_w}{\alpha_{w0} h_{w0}};\ L_w = \frac{G_w C_w}{\alpha_{w0} h_{w0}};$$

$$\beta = T_m S + \varepsilon^* + 1;\ T_m = \frac{g_m C_m}{\alpha_{w0} h_{w0}};\ \varepsilon^* = (1 - L_w^*);\ L_w^* = \frac{1}{L_w + 1};$$

$$\gamma_1 = \frac{(T_w S + 1)\beta - 1}{\beta};\ L_w^* = \frac{1}{L_w + 1};\ \xi = \frac{z}{L_w},$$

where n is the rotational speed of the electric motor of the circulating pump, rpm; α is the coefficient of heat transfer, kW/(m^2·K); t is the temperature of local water, K; f_s is the bypass for local water, m^2; C is the specific heat capacity, KJ/(kg·K); G is the flow rate of a substance, kg/s; ρ is the density of a substance, kg/m^3; g is the specific mass of a substance, kg/m; h is specific surface, m^2/m; z is the spatial coordinate of the heater's length, metres; T_w, T_m are the time constants, characterizing the thermal accumulating capacity of local water, metal, seconds; S is the parameter for a Laplace transformation; $S = \omega j$; ω is frequency, 1/s. Indices: 0 – original stationary mode; 1 – input to an electric system; w – local water; m – metallic wall.

A valid part of the transfer function (3) was determined concerning a change in the capacity of a rechargeable battery:

$$O_1(\omega) = \frac{(L_1 A_1) + (M_1 B_1) K_w K_{ec} (1 - L_w^{**})}{(A_1^2 + B_1^2)}. \tag{6}$$

The structure of coefficient K_w includes the temperature of the separating wall θ:

$$\theta = (\alpha_w (t_1 + t_2)/2) + A(t_1 + t_2)/2)/(\alpha_w + A), \tag{7}$$

where t_1 and t_2 are the local water temperature at the inlet and outlet of the thermoelectric battery, K, respectively; α is the heat transfer coefficient, kW/(m^2·K). Index: w – local water.

$$A = 1/(\delta_m/\lambda_m + 1/\alpha_w), \tag{8}$$

where δ is the thickness of a heater's wall, metres; λ is the thermal conductivity of a metal of a heater's wall, kW/(m·K); Indices: w – local water; m – metallic wall of a heater.

To use the valid part $O_1(\omega)$, the following coefficients were derived:

$$A_1 = \varepsilon^* - T_w T_m \omega^2;\ A_2 = \varepsilon^* + 1; \tag{9}$$

$$B_1 = T_w\varepsilon^*\omega + T_w\omega + T_m\omega; \tag{10}$$

$$B_2 = T_m\omega;\ C_1 = \frac{A_1A_2 + B_1B_2}{A_2^{\ 2} + B_2^{\ 2}};\ D_1 = \frac{A_2B_1 - A_1B_2}{A_2^{\ 2} + B_2^{\ 2}}; \tag{11}$$

$$L_1 = 1 - e^{-\xi C_1}\cos(-\xi D_1);\ M_1 = -e^{-\xi C_1}\sin(-\xi D_1). \tag{12}$$

The valid part of the transfer function (4) has been singled out concerning a change in the rotational speed of the electric motor of the circulating pump:

$$O_2(\omega) = K_f\chi_t L_w(C_1L_1) - (D_1M_1). \tag{13}$$

To use the valid part $O_2(\omega)$, the following coefficients were derived:

$$A_1 = -T_m\omega^2;\ A_2 = \varepsilon^* - T_wT_m\omega^2;\ B_1 = (\varepsilon^* + 1)\omega; \tag{14}$$

$$B_2 = T_w\varepsilon^*\omega + T_w\omega + T_m\omega + \varepsilon^*, \tag{15}$$

$$C_1 = \frac{A_1A_2 + B_1B_2}{A_2^{\ 2} + B_2^{\ 2}};\ D_1 = \frac{A_2B_1 - A_1B_2}{A_2^{\ 2} + B_2^{\ 2}}; \tag{16}$$

$$L_1 = 1 - e^{-\xi C_1}\cos(-\xi D_1);\ M_1 = -e^{-\xi C_1}\sin(-\xi D_1). \tag{17}$$

The valid part of the transfer function (6) was singled out concerning the estimate of the change in the consumption of local water:

$$O_3(\omega) = \chi_t L_w(C_1L_1) - (D_1M_1). \tag{18}$$

To use the valid part $O_3(\omega)$, the following coefficients were derived:

$$A_1 = -T_m\omega^2;\ A_2 = \varepsilon^* - T_wT_m\omega^2;\ B_1 = (\varepsilon^* + 1)\omega; \tag{19}$$

$$B_2 = T_w \varepsilon^* \omega + T_w \omega + T_m \omega + \varepsilon^*, \tag{20}$$

$$C_1 = \frac{A_1 A_2 + B_1 B_2}{A_2^2 + B_2^2}; \quad D_1 = \frac{A_2 B_1 - A_1 B_2}{A_2^2 + B_2^2}; \tag{21}$$

$$L_1 = 1 - e^{-\xi C_1} \cos(-\xi D_1); \quad M_1 = -e^{-\xi C_1} \sin(-\xi D_1). \tag{22}$$

Transfer functions (3) to (6), derived from using the operator method for solving a system of non-linear differential equations maintain the parameter for a Laplace transform – $S(S = \omega j)$, where ω is frequency, 1/s.

To make a transition from the frequency domain to the time domain, the valid parts of (6), (13), and (18) were singled out, derived from the mathematical treatment of transfer functions. These parts are part of integrals (23) to (25), providing a possibility to derive the dynamic characteristics of change in the capacity of a rechargeable battery, the rotational speed of the electric motor of the circulating pump, local water consumption, respectively, using the inverse Fourier transform.

$$CE(\tau) = \frac{1}{2\pi} \int_0^\infty O_1(\omega) \sin(\tau\omega / \omega) d\omega. \tag{23}$$

$$n(\tau) = G_w(\tau, z) K_f(\tau) = \frac{1}{2\pi} \int_0^\infty O_2(\omega) \sin(\tau\omega / \omega) \mathrm{d}\omega. \tag{24}$$

$$G_w(\tau, z) = \frac{1}{2\pi} \int_0^\infty O_3(\omega) \sin(\tau\omega / \omega) \mathrm{d}\omega, \tag{25}$$

where CE is the capacity of a rechargeable battery, ampere-hours; n is the number of rotations of the circulating pump's electric motor; rpm. G_w is the flow rate of local water, kg/s.

Maintaining Voltage within the Distribution System Based on a Prediction of Changes in the Battery Capacity

According to formulas (1) to (6), integrated mathematical modelling of a wind-solar electric system has been performed using the designed structural diagram (Figure 6.2).

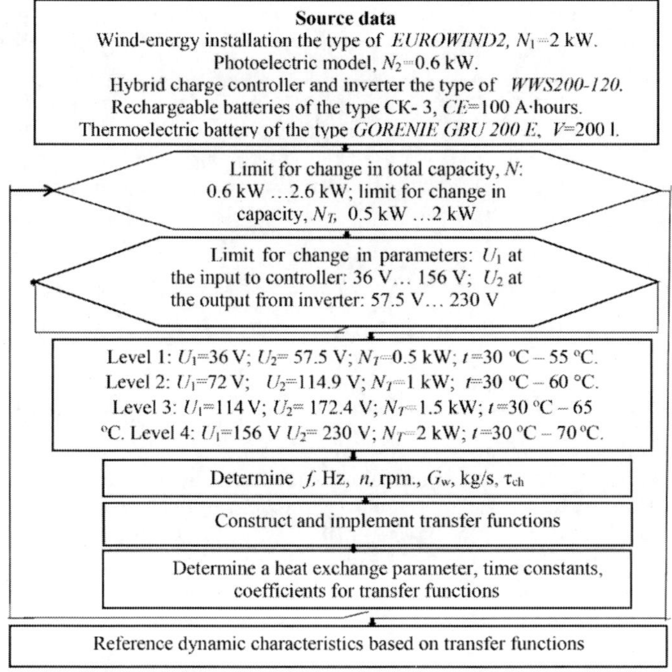

Figure. 6.2. Structural diagram of integrated mathematical modelling of a wind-solar electric system: N, N_1, N_2, N_T – total power of a wind-solar electric system, power of a wind energy installation, photoelectric module, thermoelectric battery, respectively, kW; CE – capacity of a rechargeable battery, ampere-hours; V – volume of a thermoelectric battery, l; U_1, U_2 – voltage at the hybrid charge controller input and at the inverter output, respectively, volts; t – temperature of local water, °C; f – current frequency, Hz; n – the number of rotations of the circulating pump's electric motor, rpm; G_w – flow rate of local water, kg/s.

According to the proposed structural diagram (Figure 6.2), Tables 6.1 and 6.2 give the results from integrated mathematical modelling of a wind-solar electric system.

Based on the proposed mathematical substantiation to maintain the operation of a wind-solar electric system (1) – (6), a structural diagram has been built (Figure 6.3) regarding control over the feasibility of a wind-solar electric system.

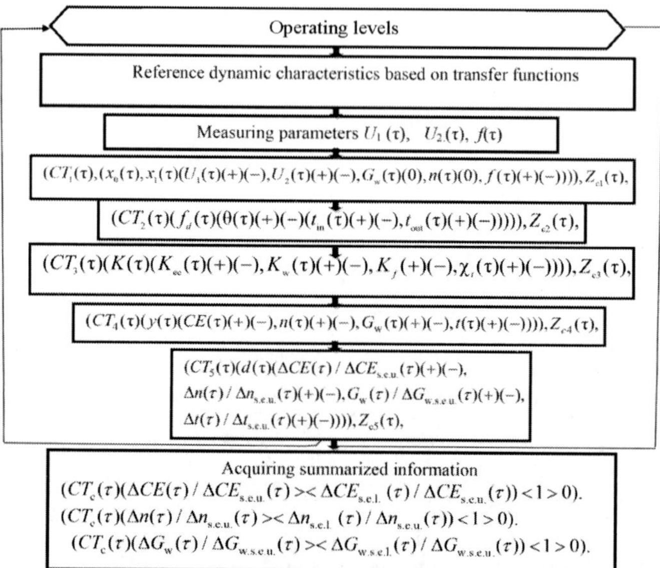

Figure 6.3. Structural diagram of control over feasibility of a wind-solar electric system: U_1, U_2 – voltage at the hybrid charge controller input and the inverter output, respectively, volts; f – current frequency, Hz; n – number of rotations of the circulating pump's electric motor, rpm; G_w – flow rate of local water, kg/s; t_{input}, t_{output}, θ – local water temperature at the inlet and outlet of a thermoelectric battery, separating wall, respectively, K; CE – capacity of a rechargeable battery, ampere-hours; CT – event control; Z – logical relationships; d – dynamic parameters; x – influences; f_d – diagnosed parameters; y – starting parameters; K – coefficients for mathematical notation; ι – time. Indices: c – operability control; s.e.u. – steady, estimated value for the parameter of the upper level of operation; s.e.l. – steady, estimated value for the parameter of the level of operation; 0,1,2 – initial stationary mode, external, internal parameters; 3 – coefficients for equations of dynamics; 4 – essential parameters that are diagnosed; 5 – dynamic parameters.

Table 6.1. Mode parameters for a wind-solar electric system

Operating levels	N, kW	N_T, kW	t,°C	G_w, kg/s	τ_{ch}, hours	U_1, V	U_2, V	f, Hz	n, rpm
Level 1	0.6	0. 5	55	0.0024	3.87	36	57.5	12.5	712.5
Level 2	1.2	1	60	0.0043	0.21	72	114.9	25	1425
Level 3	1.9	1.5	65	0.0060	0.17	114	172.4	37.5	2137.5
Level 4	2.6	2	70	0.0073	0.098	156	230	50	2850

Note: N, N_T – total power of a wind-solar electric system, thermoelectric battery, respectively, kW; t – local water temperature at the thermoelectric battery outlet, °C; τ_{ch} – charge duration of a thermoelectric battery; G_w – flow rate of local water, kg/s; U_1 – voltage at the hybrid charge controller input, volts; U_2 – voltage at the inverter output, volts; f – current frequency, Hz; n – number of rotations of the circulating pump's electric motor, rpm.

Table 6.2. Values for a heat exchange parameter, for time constants and coefficients for the mathematical models of dynamics

Operating levels	α_w, W/(m^2·K)	$T_{w,}$ s	T_m, s	L_w, m	L_w^*	ε^*	ζ
Level 1	507.2	97.85	6.65	0.1316	0.8837	01116	3.32
Level 2	564.7	87.54	5.97	0.2117	0.8253	0.1677	2.08
Level 3	595	82.53	5.67	0.2804	0.7810	0.2102	1.57
Level 4	636.8	78.82	5.29	0.3187	0.7583	0.2320	1.38

Note: α_w – coefficient of convective heat transfer from an electric heater to local water, W/(m^2·K).

An Integrated System to Maintain the Operation of a Wind-Solar Electric System at the Decision-Making Level

Based on the proposed mathematical substantiation (1)–(6), a structural diagram has been built (Figure 6.4) to maintain the operation of a wind-solar electric system based on maintaining the capacity of a rechargeable battery.

Results and Discussion

The Smart Grid System for Maintaining the Operation of the Wind-Solar Electric System at the Decision-Making Level

An integrated comprehensive system to support the operation of a wind-solar electric system has been devised (Table 6.3), which is based on predicting a change in the capacity of a rechargeable battery using continuous measurement of voltages at the hybrid charge controller input, the inverter output, and current frequency.

Making advance decisions about changing the power of a thermoelectric battery makes it possible, while maintaining the capacity of a rechargeable battery, to enable a change in the local water temperature based on changing the number of rotations of the circulating pump's electric motor in order to change water consumption.

The integrated Smart Grid system for operational maintenance of the wind-solar electric system (Table 6.3) provides an opportunity to coordinate electric power production and consumption.

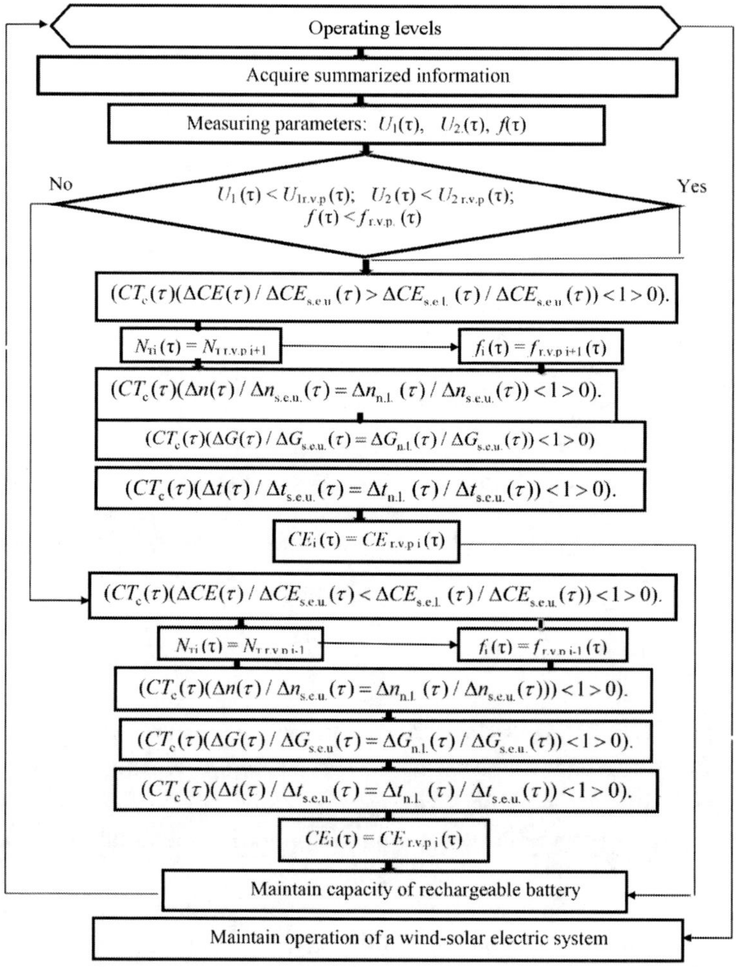

Figure 6.4. Structural diagram for maintaining the operation of a wind-solar electric system: U_1, U_2 – voltage at the hybrid charge controller input and the inverter output, respectively, volts; f – current frequency, Hz; CE – capacity of a rechargeable battery, ampere-hours; N_T – thermoelectric battery power, kW; n – number of rotations of the circulating pump's electric motor, rpm; G – flow rate of local water, kg/s; t – temperature of local water, K; ι – time. Indices: i – number of operating levels; e – reference value for parameter; s.e.u. – steady, estimated value for the parameter of the upper operating level; s.e.l. – steady, estimated value for the parameter of the operating level; n.l. – new operating level.

Table 6.3. Integrated system to maintain the capacity of a rechargeable battery

Time, τ, 10^3 s	Change in parameters	$\Delta CE(\tau)/\Delta CE_{s,e,u,}(\tau)$	$\Delta n(\tau)/\Delta n_{s,e,u,}(\tau)$	$\Delta G_w(\tau)/\Delta G_{w\,s,e,u,}(\tau)$	$CE(\tau)$, A·h	$n(\tau)$, rpm	$G_w(\tau)$, kg/s
0	Charge $U_1 = 12$ V; $U_2 = 57.5$ V; $f = 2.5$ Hz; $t = 55$°C	0.0473	0.0327	0.3252	4.73	712.5	0.0024
3	Charge $U_1 = 18$ V; $U_2 = 57.5$ V; $f = 2.5$ Hz; $t = 55$°C	0.0817	0.0654	0.3614	8.17	782.4	0.0027
6	Charge $U_1 = 24$ V; $U_2 = 57.5$ V; $f = 7.5$ Hz; $t = 55$°C	0.1285	0.0982	0.4068	12.85	852.5	0.0030
9	Charge $U_1 = 30$ V; $U_2 = 57.5$ V; $f = 10$ Hz; $t = 55$°C	0.1946	0.1309	0.4646	19.46	922.4	0.0034
12	Charge $U_1 = 36$ V; $U_2 = 57.5$ V; $f = 12.5$ Hz; $t = 55$ °C	0.2987	0.1636	0.5420	29.87	992.3	0.0040
15	Decision to discharge $U_1 = 42$ V; $U_2 = 57.5$ V; $f = 12.5$ Hz; $t = 55$°C	0.4865	0.1998	0.6505	48.65	1069.7	0.0045
18	Discharge $f = 20$ Hz; $U_1 = 42$ V; $U_2 = 92$ V; $t = 60$°C	0.3198	0.319	0.7	31.98	1325.8	0.0047
21	Charge $U_1 = 54$ V; $U_2 = 114.9$ V; $f = 25$ Hz; $t = 60$°C	0.3321	0.3996	0.7447	33.21	1496.8	0.0049
24	Charge $U_1 = 72$ V; $U_2 = 114.9$ V; $f = 25$ Hz; $t = 60$°C	0.6285	0.4351	0.8183	62.85	1496.8	0.0053
27	Decision to discharge $U_1 = 78$ V; $U_2 =$ 114.9 V; $f = 25$ Hz; $t = 60$°C	0.7914	0.4351	0.8410	79.14	1496.8	0.0054
30	Discharge $f = 30$ Hz; $U_1 = 78$ V; $U_2 = 137.9$V; $t = 65$°C	0.5785	0.5222	0.8651	57.85	1758.4	0.0055
33	Charge $U_1 = 114$ V; $U_2 = 172.4$ V $f = 37.5$ Hz; $t = 65$°C	0.8676	0.6526	0.8894	86.76	2,037.1	0.0056
36	Decision to discharge $U_1 = 120$ V; $U_2 = 172.4$ V; $f = 37.5$ Hz; $t = 65$°C	1	0.7548	0.9096	100	2,255.5	0.0057
39	Discharge $f = 40$ Hz; $U_1 = 120$ V; $U_2 = 183.9$ V; $t = 70$°C	0.8856	0.8053	0.93	88.56	2363.4	0.0058
42	Charge $U_1 = 150$ V; $U_2 = 230$ V; $f = 50$ Hz; $t = 70$°C	0.8894	1	0.9530	88.94	2780.6	0.0059
45	Charge $U_1 = 156$ V; $U_2 = 230$ V; $f = 50$ Hz; $t = 70$°C	1	1	1	100	2850	0.0073

Note: U_1, U_2 – voltage at the hybrid charge controller input and the inverter output, respectively, volts; f – current frequency, Hz; t – local water temperature at the thermoelectric battery outlet, °C; CE – capacity of a rechargeable battery, ampere-hours; n – number of rotations of the circulating pump's electric motor, rpm; G_w – flow rate of local water, kg/s; τ – time, seconds. Index: s.e.u. – steady, estimated value for the parameter of the upper operating level.

Coordination of Electric Power Production and Consumption Based on Voltage Maintenance within the Distribution System

The capacity of a rechargeable battery over the established period of time is determined from:

$$CE_i(\tau) = CE_i(\Delta CE_i(\tau) / \Delta CE_{\text{с.р.в.}}(\tau), \tag{26}$$

where CE – capacity of a rechargeable battery, ampere-hours; τ – time, seconds. Index: *s.e.u.* – steady, estimated value for the parameter of the upper operating level; i – number of operating levels of a wind-solar electric system.

The rotational speed of the electric motor of the circulating pump over the established period is determined from:

$$\begin{aligned} n_{i+1}(\tau) = n_i + ((\Delta n_{i+1}(\tau) / \Delta n_{s.e.u.}(\tau) - \\ -\Delta n_i(\tau) / \Delta n_{s.e.u.}(\tau))(n_2 - n_1)), \end{aligned} \tag{27}$$

where n is the number of rotations of the circulating pump's electric motor, rpm; n_1, n_2 – initial and final number of rotations of the circulating pump's electric motor, rpm, respectively; i – number of operating levels of a wind-solar electric system; τ – time, seconds. Index: s.e.u. – steady, estimated value for the parameter of the upper operating level.

The local water flow rate over the established period is determined from:

$$\begin{aligned} G_{w_i+1}(\tau) = G_{wi} + ((\Delta G_{wi+1}(\tau) / \Delta G_{w\ s.e.u.}(\tau) - \\ -\Delta G_{wi}(\tau) / \Delta G_{w\ s.e.u.}(\tau))(G_{w_2} - G_{w1})), \end{aligned} \tag{28}$$

where G_w – local water flow rate, kg/s; G_{w1}, G_{w2} – initial and final values for local water consumption, kg/s, respectively (Table 5); i – number of operating levels of a wind-solar electric system; τ – time, seconds. Index: s.e.u. – steady, estimated value for the parameter of the upper operating level; i – number of operating levels of a wind-solar electric system. For example, over the period of $15 \cdot 10^3$ s (4.17 hours) the absolute values for capacity of a rechargeable battery, the number of rotations of the circulating pump's electric motor, local water consumption, applying formulae (26) to (28), are equal to:

$$48.65\ \text{A}\cdot\text{h} = (0.4865)100\ \text{A}\cdot\text{h}$$

$$1{,}069.7\ \text{rpm} = 992.3\ \text{rpm} + (0.1998 - 0.1636)(2{,}850\ \text{rpm} - 712.5\ \text{rpm}).$$

$$0.0045\ \text{kg/s} = 0.0040\ \text{kg/s} + (0.6505 - 0.5420)(0.0073\ \text{kg/s} - 0.0024\ \text{kg/s}).$$

The graphical dependence of change in the capacity of a rechargeable battery on change in the power of a thermoelectric battery is shown in Figure 6.5.

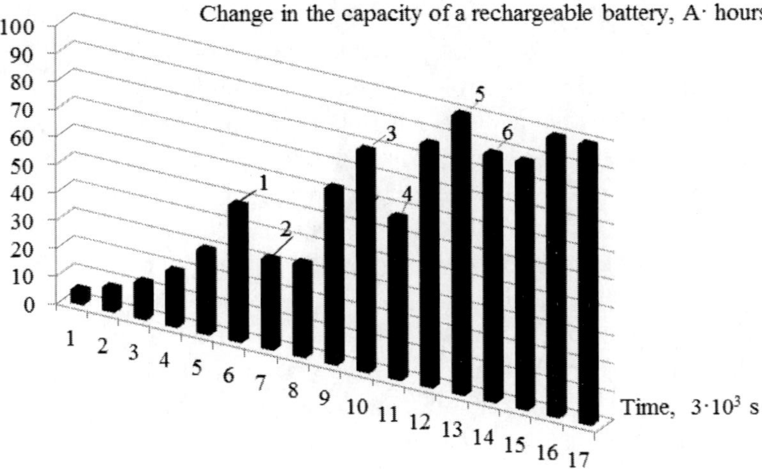

Figure 6.5. Maintaining the operation of a wind-solar electric system, where 1, 3, and 5 are points when decisions were made to change the power of a thermoelectric battery, while 2, 4, and 6 – decisions to discharge to maintain a change in the capacity of a rechargeable battery.

Thus, for example, over time of $15\cdot10^3$ s (4.17 hours) when changing voltage at the hybrid charge controller input to the level of 42 V, while predicting an increase in the capacity of a rechargeable battery to 20%, it is necessary to make an advance decision about increasing the capacity of a thermoelectric battery based on a change in current frequency from 12.5 Hz to 20 Hz. Increasing the number of rotations of the circulating pump's electric motor to the level of 1,328 rpm makes enables increasing the local water consumption to the level of 0.0047 kg/s and increasing the local water temperature up to 60°C. Implementing these measures will enable maintaining a 20% change in the capacity of a rechargeable battery, thereby maintaining the operation of a wind-solar electric system. An integrated system to maintain the operation of a wind-solar electric system has been built, based on adjusting the generation and consumption of energy in terms of energy efficiency. Typically, a hybrid charge controller included in a wind-solar electric system maintains the charge of a rechargeable battery by using a thermoelectric battery as a non-regulated ballast. Resetting excess energy to the ballast when using the MPPT function of the controller leads to unrecoverable losses of electrical energy, which does not make it possible to ensure an appropriate level of capacity for the charge of a rechargeable battery. Moreover, the use of a thermoelectric battery as a ballast eliminates the need to maintain the operation of a wind energy installation by using the accumulation of heat in order to regulate the power of a wind turbine. Failure to account for this property of a thermoelectric battery could lead to the acceleration of the wind turbine at considerable wind speed and to its malfunctioning. It is known that a thermoelectric battery is controlled in line with the thermostat principle, that is, when establishing the required temperature of heated local water, the thermoelectric battery is disconnected from power. Not using a change in the local water flow rate during the period of thermoelectric battery charge, when changing capacity, prolongs the duration of charge and leads to considerable expenditures on electricity. A technique to overcome these difficulties has been proposed. It is the thermoelectric battery that must become the main centre of adjusting a change in the total power of a wind-solar electric system to power consumption by redistributing the accumulated heat and electric energy in terms of consumption. It has been proposed to predict a change in the capacity of a rechargeable battery when measuring the total voltage at the hybrid charge controller input and voltage at the inverter output to estimate the ratio of electrical energy generation and consumption when measuring the current frequency. This ratio is part of the coefficient K_{ce} for a transfer function aimed at predicting a change in the capacity of a rechargeable battery, which could act as a unifying element in the integrated mathematical and logical modelling as part of a technological system to maintain the operation of a wind-solar electric system. Making advance decisions on changing the power of a thermoelectric battery makes it possible, while maintaining the capacity of the rechargeable battery, to ensure a change in the temperature of heated local water, based on changing the number of rotations of the circulating pump's electric motor in terms of water flow rate, thereby reducing the charge duration by up to 30%. generation of electric power to efficient consumption.

Chapter 7

Smart Grid Technology for Maintaining the Operation of a Wind-Solar Electric System Network

Abstract

Integrated Smart Grid Systems for harmonization of production and consumption of electric power of the heat pump power supply and hot water power supply are being developed. The integrated dynamic subsystem of the wind-solar electric system includes the following components: the electric network, wind turbine, photovoltaic solar panels, hybrid solar collectors, grid inverter, heat pump, two-section storage tank, upper section for hot water supply, lower section a low-grade energy source, frequency converter. Integrated systems based on predicting changes in the power factor and local water temperature by measuring voltage from hybrid solar collectors at the grid inverter input, voltage at the frequency converter output, as well as current frequency. The adoption of advance decisions to maintain the local water temperature by changing the power of the electric motor of the heat pump compressors and the electric motor of the circulation pump is based on establishing the ratio of the voltages at the grid inverter input and the frequency converter output. The power factor of the wind-solar electric system is maintained. The complex mathematical and logical modelling of the wind-solar electric system, based on the mathematical substantiation of the architecture of the wind-solar electric system and mathematical substantiation of the operational maintenance of the wind-solar electric system, is performed. Time constants and coefficients of the mathematical models of dynamics regarding the estimation of a change in the power factor of the system, and the local water temperature are predicted by measuring voltage from hybrid solar collectors at the grid inverter input, voltage at the frequency converter output and current frequency. Functional estimation of a change in the power factor of the wind-solar electric system is in the range of 58–98%, and local water temperature is in the range of 30–55°C. Determining final functional information provides an opportunity to make advance decisions on a change in the operation of the electric motor in the heat pump compressor and the electric motor in the circulation pump to prevent the peak load on the power system and maintain voltage when the heat pump power supply and hot water supply are connected.

Keywords: wind-solar power system, hybrid solar collectors

Introduction

The distributed generation of electricity using renewable sources requires intelligent systems for managing electricity flows and consumption. Smart Grid technologies, demand management systems and energy storage are new components for the integration of distributed energy generation in the energy system. An urgent further development in this direction is predicting changes in the power factor, local water temperature of heat pump power supply and hot water supply as part of a wind-solar electric system network using hybrid solar collectors. In terms of connection to intelligent control systems, the author, Chaikovskaya, E. (2017) proposes to predict voltage changes when measuring the temperature of the electrolyte in the volume of batteries. Energy-saving technology was developed for the operation of a storage battery; it prevents overcharging and enables discharging based on the coordination of electrochemical and diffusion processes of discharging and charging. The author's work (2019) is devoted to forecasting changes in parameters for connecting to Smart Grid technologies, which presents an integrated system for supporting the operation of a wind-solar electric system based on predicting changes in battery capacity. The voltage is measured at the hybrid charge controller input and the inverter output. The adoption of advance decisions to change the power of the thermoelectric accumulator is based on establishing the ratio of the voltage at the hybrid charge controller input and the inverter output when measuring the current frequency. The change in the flow rate and temperature of the heated water is provided, reducing the charge period to 30%, based on the change in the number of revolutions of the circulation pump electric motor. Thus, in the work (Yinan, L., Wentao, Y., & Ping, H., Chang, Ch., Xiaonan, W., 2019) it is proposed to control the consumption of electrical energy by exchanging information in real-time. The work (Saad, A., Samy, F., & Osama, M., 2019) is devoted to the introduction of stochastic optimization of distributed generation of electrical energy using fuzzy logic. The work (Perera, A., Vahid M., & Wickramasinghe, P., Scartezzini, J., 2019) proposes a cyber-physical control system for the distributed generation of electrical energy, based on the theory of a consensus protocol. the work (Davye, M., Daranith, & Ch., Dae-Hyun, Ch., 2020) proposes an intelligent converter for voltage regulation by absorbing or supplying reactive power. The work (Xiqiao, L., Yukun, L., & Xianhong, B., 2019) presents a model for prioritizing sensitivity to changes in data based on accurate measurement of electrical energy consumption.

Methodological and Mathematical Substantiation

One of the main properties of energy systems is the mandatory exchange of substance, energy, and information with the environment. Thus, the wind-solar electric system is an open integrated dynamic system, the operation of which requires predicting changes in the power factor, and local water temperature when measuring voltage from hybrid solar collectors at the grid inverter input, voltage at the frequency converter output and voltage frequency. The dynamic characteristics of the wind-solar electric system with sufficient accuracy for practice can be described by a finite set of parameters for changes in time, and the spatial coordinate that coincides with the direction of flow of the medium, such as changes in power factor and local water temperature. Therefore, the dynamic description of the wind-solar electric system most fully and multifacetedly characterizes its operation. Thus, it is possible to determine that a real wind-solar electric system is a dynamic system, the mathematical model of which reflects the properties of the transformation of influences, i.e., its dynamic properties. Because the wind-solar electric system reflects the dynamic peculiarities due to the nature of reactions to influences, the operational maintenance of the wind-solar electric system should be part of such a technological system, which is based on a dynamic system.

Based on the methodological, mathematical, and logical substantiation of the technological systems (Chapter 1) the architecture, mathematical substantiation of the architecture (1), and mathematical substantiation of operational maintenance (2) of the Smart Grid wind-solar electric system are proposed (Figure 7.1).

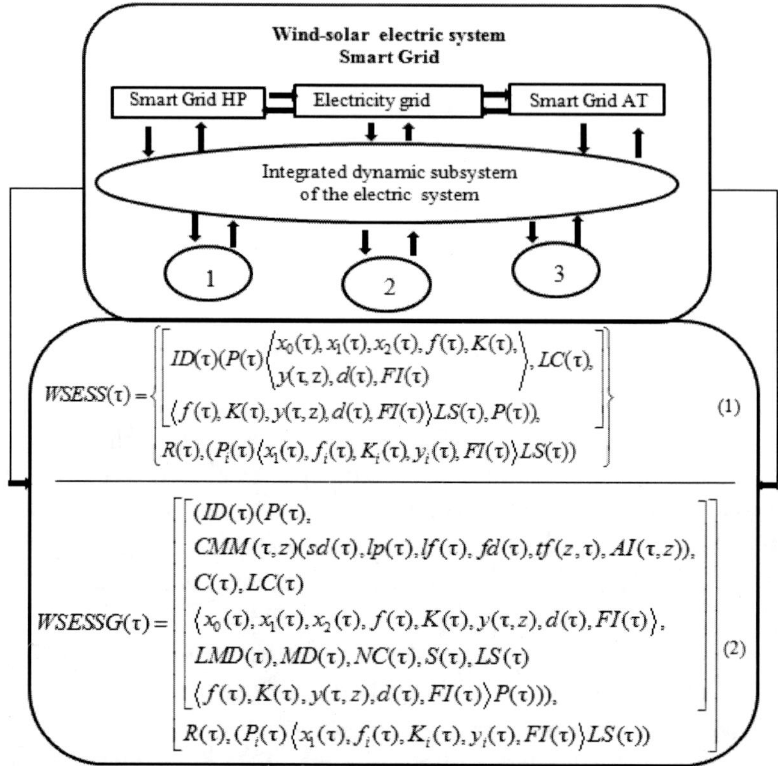

$$WSESS(\tau)=\left\{\begin{matrix}\left[\begin{matrix}ID(\tau)(P(\tau)\left\langle\begin{matrix}x_0(\tau),x_1(\tau),x_2(\tau),f(\tau),K(\tau),\\ y(\tau,z),d(\tau),FI(\tau)\end{matrix}\right\rangle,LC(\tau),\\ \langle f(\tau),K(\tau),y(\tau,z),d(\tau),FI(\tau)\rangle LS(\tau),P(\tau)),\end{matrix}\right]\\ R(\tau),(P_i(\tau)\langle x_1(\tau),f_i(\tau),K_i(\tau),y_i(\tau),FI(\tau)\rangle LS(\tau))\end{matrix}\right\} \quad (1)$$

$$WSESSG(\tau)=\left|\begin{matrix}\left[\begin{matrix}(ID(\tau)(P(\tau),\\ CMM(\tau,z)(sd(\tau),lp(\tau),lf(\tau),fd(\tau),tf(z,\tau),AI(\tau,z)),\\ C(\tau),LC(\tau)\\ \langle x_0(\tau),x_1(\tau),x_2(\tau),f(\tau),K(\tau),y(\tau,z),d(\tau),FI(\tau)\rangle,\\ LMD(\tau),MD(\tau),NC(\tau),S(\tau),LS(\tau)\\ \langle f(\tau),K(\tau),y(\tau,z),d(\tau),FI(\tau)\rangle P(\tau))),\end{matrix}\right]\\ R(\tau),(P_i(\tau)\langle x_1(\tau),f_i(\tau),K_i(\tau),y_i(\tau),FI(\tau)\rangle LS(\tau))\end{matrix}\right| \quad (2)$$

Figure 7.1. Smart Grid wind-solar electric system: the architecture: HP – heat pump; AT – accumulator tank; 1 – charging unit; 2 – discharging unit; 3 – unit for functional efficiency evaluation. Mathematical substantiation of the architecture (1). Mathematical substantiation of operational maintenance (2).

A wind-solar electric system is a dynamic system, the operation of which is the reproduction of a change in external, internal influences and initial conditions, for example, changes in solar radiation, wind speed, change in consumption of electrical energy and heat. Therefore, when designing a wind-solar electric system, an integrated dynamic subsystem is laid down in its base (Figure 7.1). The integrated dynamic subsystem includes the following components: mains, wind turbine, photovoltaic solar panels, hybrid solar collectors, grid inverter, heat pump, two-section storage tank, upper section for hot water supply, lower section as a low-grade energy source, and frequency converter. When the system design is represented as the organization of a complex system, it is expanded by building up the dynamic subsystem blocks that forecast the process components around its base. Other components of the wind-solar electric system include the units for charging, discharging, and functional efficiency estimation in a coordinated interaction with the dynamic subsystem (Figure 7.1). The mathematical substantiation of the architecture of the wind-solar electric system Smart (1), (Figure 7.1), based on the methodology of the mathematical description of dynamics of power systems, the method of the graph of cause-effect relations (Chapter 1) is proposed.

Where $WSESSG(\tau)$ – Smart Grid wind-solar electric system; τ – time, seconds; $ID(\tau)$ – integrated dynamic subsystem (mains, wind turbine, photovoltaic solar panels, hybrid solar collectors, grid inverter, heat pump, two-section storage tank, upper section for hot water supply, lower section as a low-grade energy source, frequency converter); $P(\tau)$ – properties of the components of the wind-solar electric system; $x(\tau)$ – impacts (change in solar radiation, wind speed, change in consumption of electrical energy and heat; $f(\tau)$ – parameters that are measured: (voltage at the grid inverter input from hybrid solar collectors, voltage at the frequency converter output, current frequencies); $K(\tau)$ – coefficients of mathematical description of the dynamics of power factor of the wind-solar electric system, the local water temperature; $y(\tau, z)$ – predicted output parameters (power factor, the local water temperature) z – spatial coordinate of the condenser axis, heat exchanger coincides with the direction of the flow of motion of the medium, metres; $d(\tau)$ – dynamic parameters of power factor change, the local water temperature change; $FI(\tau)$ – functional resulting information on decision-making; $LC(\tau)$ – logical relations regarding the operability control of the wind-solar electric system; $LS(\tau)$ – logical relations regarding the identification of the state of the wind-solar electric system; $R(\tau)$ – logical relations in $WSESSG(\tau)$ to confirm the correctness of decisions made from the units of the wind-solar system. Indices: i – the number of elements of the wind-solar system; 0,1,2 – initial stationary mode, external, internal nature of impacts.

The mathematical substantiation of operational maintenance of the Smart Grid photoelectric charging station (2), (Figure 7.1), based on the methodology of the mathematical description of dynamics of power systems, the method of the graph of cause-effect relations (Chater 1) is proposed. The basis of the proposed rationale is the mathematical description of the architecture of the wind-solar electric system Smart (1), (Figure 7.1). Prediction of changes in the power factor, local water temperature when measuring voltage from hybrid solar collectors at the grid inverter input, voltage at the frequency converter output and current frequency enables making advance decisions to maintain the local water temperature by changing the power of the electric motor of the heat pump compressors and electric motor of the circulation pump based on establishing the ratio of the voltage at the grid inverter input and the voltage at the frequency converter output are measured. The power factor of the wind-solar electric system is maintained. The mathematical substantiation of operational maintenance of the Smart Grid wind-solar system (2) is proposed (Figure 7.1).

Where *WSESSG*(τ) – operational maintenance of the Smart Grid wind-solar electric system; τ – time, seconds; *ID*(τ) – integrated dynamic subsystem (mains, wind turbine, photovoltaic solar panels, hybrid solar collectors, grid inverter, heat pump, two-section storage tank, upper section for hot water supply, lower section as a low-grade energy source, frequency converter). *P*(τ) – the properties of the elements of the integrated dynamic subsystem, units of the wind-solar electric system; *CMM*(τ,*z*) – complex mathematical modelling of the dynamics of changes in power factor, the local water temperature; *sd*(τ) – the input data (the wind-solar electric system network with a capacity of 10 kW includes the following components: EUROWIND2 type (Ukraine) – a wind power plant; ATMOSFERA-F2PV (Ukraine) – photovoltaic solar panels and hybrid solar panels. The heat pump system – Commotherm Hybrid Tower WW, Split DeLuxe (Austria) with a heating capacity of 5.7 kW is equipped with a two-section storage tank, the lower section of which has a volume of 200 litres, used as a low-potential energy source connected to hybrid solar collectors. The upper section of it has a volume of 300 litres, used as a heat accumulator connected to hybrid solar collectors; *lp*(τ) – the boundary change in parameters (voltage from hybrid solar collectors at the grid inverter input, voltage at the frequency converter output and current frequency; *lf*(τ) – the levels of operation of the wind-solar electric system; *fd*(τ) – the obtained parameters (mode parameters of the wind-solar electric system); *tf*(τ,*z*) – the transfer function of predicted parameters: in the power factor, local water temperature; *AI*(τ,*z*) – reference information for assessing changes in power factor, the local water temperature; *C*(τ) – the control of operability of the wind-solar electric system; *LC*(τ) – the logical relations of the operability control of the wind-solar electric system; *x*(τ) – impacts (change in solar radiation, wind speed, change in consumption of electrical energy and heat; *f*(τ) – the measured parameters: (voltage at the grid inverter input from hybrid solar collectors, voltage at the frequency converter output, current frequencies); *K*(τ) – coefficients of mathematical description of the dynamics of power factor of the wind-solar electric system, the local water temperature;; *y*(τ, *z*) –

predicted output parameters (power factor, the local water temperature); z – spatial coordinate of the condenser axis, heat exchanger coincides with the direction of the medium flow, metres; $d(\tau)$ – dynamic parameters of power factor change, the local water temperature change; $FI(\tau)$ – functional resulting information on decision-making; $LMD(\tau)$ – the logical relations of decision-making; $MD(\tau)$ – decision-making; $NC(\tau)$ – the new conditions of the wind-solar electric system; $S(\tau)$ – the identification of the state of the wind-solar electric system; $LS(\tau)$ – the logical relations of identification of the state of the wind-solar electric system; $R(\tau)$ – the logical relations between the dynamic subsystem and units for charging, discharging, and functional estimation of efficiency that belong to the wind-solar electric system. Indices: i – the number of elements of $WSESSG(\tau)$; 0,1,2 – the initial, external, and internal character of influences.

Mathematical substantiation of the architecture of the photoelectric charging station Smart (1) and mathematical substantiation of operational maintenance of the Smart Grid photoelectric charging station (2) (Figure 7.1) enable maintaining the operation of the photoelectric charging station using the following actions:

- operability control ($C(\tau)$) of the dynamic subsystem based on complex mathematical ($CMM(\tau,z)$) and logical ($LC(\tau)$) modelling regarding obtaining standard ($AI(\tau,z)$) estimate of a change in the power factor of a change in the local water temperature;
- operability control ($C(\tau)$) of the dynamic system based on complex mathematical ($CMM(\tau,z)$) and logical ($LC(\tau)$) modelling regarding the obtaining functional ($FI(\tau)$) estimate of a change in the power factor of the wind-solar electric system of a change in the local water temperature;
- decision-making ($MD(\tau)$) with the use of the functional resulting information ($FI(\tau)$), obtained based on logical modelling ($LMD(\tau)$) to maintain the local water temperature by changing the power of the electric motor of the heat pump compressors and electric motor of the circulation pump based on establishing the ratio of the voltage at the grid inverter input and the voltage at the frequency converter output are measured;
- identification ($S(\tau)$) of the new operational conditions of the wind-solar electric system ($NC(\tau)$) based on logical modelling ($LS(\tau)$) as a part of the dynamic subsystem and confirmation of new operating

conditions based on logical modelling ($R(\tau)$) from the units of the wind-solar electric system.

Complex Mathematical Modelling of Heat Pump Power Supply and Hot Water Supply Using Hybrid Solar Collectors

According to formulas (1) and (2), the prediction of a change in the power factor of the wind-solar electric system and the local water temperature was proposed. The voltage at the inlet to the grid inverter from hybrid solar collectors, and the outlet from the frequency converter and current frequency is measured. The transfer function for the "power factor of the wind-solar electric system – voltage at the inlet to the grid inverter" relation is complex. A change in power factor and the local water temperature are estimated. A change in the local water temperature is estimated both over time and along the spatial coordinate of the condenser axis, heat exchanger coincides with the direction of the flow of motion of the medium. The transfer function for the "power factor of the wind-solar electric system – voltage at the inlet to the grid inverter" relation, which was obtained as a result of solving a system of nonlinear differential equations, is as follows:

$$W_{PF-U_1} = \frac{K_{pf} K_t \varepsilon \left(1 - L_e^*\right)}{(T_w S + 1)\beta - 1}\left(1 - e^{-\gamma\xi}\right), \tag{3}$$

where

$$K_{pf} = \frac{I(U_1 - U_2)}{(N_e)};\ K_t = \frac{m(\theta_o - \sigma_0)}{G_{i0}};\ \varepsilon = \frac{\alpha_{e0} h_{e0}}{\alpha_{i0} h_{i0}};$$

$$L_e^* = \frac{1}{L_e + 1};\ L_e = \frac{G_e C_e}{\alpha_{e0} h_{e0}};\ T_w = \frac{g_w C_w}{\alpha_{i0} h_{i0}};\ \beta = T_m S + \varepsilon^* + 1;$$

$$\varepsilon^* = \varepsilon(1 - L_e^*);\ \gamma = \frac{(T_w S + 1)\beta - 1}{\beta};\ \xi = \frac{z}{L_i};\ L_i = \frac{G_i C_i}{\alpha_{i0} h_{i0}}.$$

where *PF* is the power factor of the wind-solar electric system; *I* – current, A; U_1 and U_2 – voltage at the grid inverter input and at the frequency converter output, respectively, volts; *Ne* – the power of the wind-solar electric system, kW; *C* is the specific thermal capacity, kJ/(kg·K); α is the heat transfer factor, kW/(m²·K); *G* is the loss of substance, kg/s; *g* is the specific weight of a substance, kg/m; *h* is the specific surface, m²/m; σ, θ are the temperature of the warming heat carrier and of the separating wall, respectively, K; *z* is spatial coordinate along the condenser axis, heat exchanger coincides with the direction of the medium flow, metres; T_w and T_m are the time constants that characterize the thermal accumulation in local water and metal, seconds; *m* is the indicator of the dependence of heat transfer factor on consumption; τ is the time, seconds; *S* is the Laplace parameter, *S* = ω*j*; ω is the frequency, 1/s. Indices: 0 – initial stationary mode; *i* – internal flow – local water; e – external flow – warming heat carrier; *m* – metal wall.

A real part of the transfer function was separated:

$$O(\omega) = \frac{(L_1 A_1) + (M_1 B_1) K_{\text{pf}} K_{\text{t}} \varepsilon (1 - L_{\text{e}}^{*})}{(A_1^{2} + B_1^{2})}. \tag{4}$$

The K_t factor includes the temperature of the separating wall θ:

$$\theta = \left(\alpha_i \frac{(\sigma_1 + \sigma_2)}{2} + A \frac{(t_1 + t_2)}{2} \right) / \left(\alpha_i + A \right), \tag{5}$$

where σ_1 and σ_2 are the temperatures of the warming heat carrier at the inlet and the outlet of the heat exchanger, K, respectively; t_1 and t_2 are the local water temperatures at the heat exchanger inlet and outlet, K, respectively; α is the heat transfer factor, kW/(m²·K). Index: *i* – local water.

$$A = \frac{1}{(\delta_{\text{m}} / \lambda_{\text{m}} + 1 / \alpha_{\text{e}})}, \tag{6}$$

where δ is the thickness of a wall of the heat exchanger, metres; λ is the thermal conductivity of the metal wall of the heat exchanger, kW/(m·K). Indices: e – warming heat carrier; *m* – metal wall of a heat exchanger.

To use the real part $O(\omega)$, the following factors were obtained:

$$A_1 = \varepsilon * - T_w T_m \omega^2; \tag{7}$$

$$A_2 = \varepsilon^* + 1; \tag{8}$$

$$B_1 = T_w \varepsilon\omega + T_w \omega + T_m \omega; \tag{9}$$

$$B_2 = T_m \omega; \tag{10}$$

$$C_1 = \frac{A_1 A_2 + B_1 B_2}{A_2^2 + B_2^2}; \tag{11}$$

$$D_1 = \frac{A_2 B_1 - A_1 B_2}{A_2^2 + B_2^2}; \tag{12}$$

$$L_1 = 1 - e^{-\xi C_1} \cos(-\xi D_1); \tag{13}$$

$$M_1 = -e^{-\xi C_1} \sin(-\xi D_1). \tag{14}$$

The transfer function (3), which was obtained based on the use of the operator method of solving the system of nonlinear differential equations, retains the Laplace transform parameter – $S(S = \omega j)$, where ω is the frequency, 1/s. To switch from the frequency area to the time area, a real part (4), obtained as a result of the mathematical treatment of the transfer function, was separated. It is this part that is included in the integrals (15) and (16), which makes it possible to obtain dynamic characteristics of a change in power factor of the wind-solar electric system, the local water temperature using the inverse Fourier transform:

$$PF(\tau) = \frac{1}{2\pi} \int_0^{\infty} K_{pf} K_t O(\omega) \sin(\tau\omega/\omega) d\omega, \tag{15}$$

$$t(\tau, z) = \frac{1}{2\pi} \int_0^{\infty} K_{pf} K_t O(\omega) \sin(\tau\omega/\omega) d\omega, \tag{16}$$

where *PF* is the power factor of the wind-solar electric system; *t* is the local water temperature, K.

So, for obtaining the reference estimation of change in the power factor of the wind-solar electric system; the local water temperature a block diagram is proposed (Figure 7.2) using, for example, the initial data of a grid-type wind-solar electric system with a power of 10 kW using hybrid solar collectors to support the operation of a heat pump power supply and hot water supply.

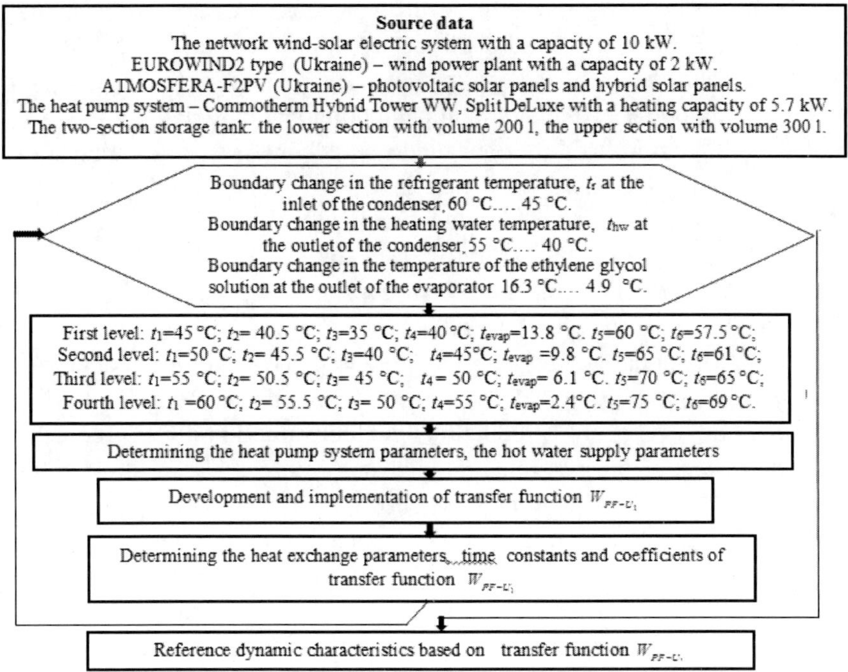

Figure 7.2. Block diagram of comprehensive mathematical modelling of the wind-solar electric system: t_1 and t_2 – the refrigerant temperature at the inlet of the condenser and the outlet of the condenser, respectively, °C; t_3 and t_4 – the local water temperature at the inlet of the condenser and the outlet of the condenser, respectively, °C; t_5 and t_6 – the temperature of the water from hybrid solar collectors at the inlet to the heat accumulator and at the outlet from the heat accumulator, respectively, °C; t_{evap} – the refrigerant evaporation temperature.

The wind-solar electric system network with a capacity of 10 kW includes the following components: EUROWIND2 type (Ukraine) – a wind power plant; ATMOSFERA-F2PV (Ukraine) – photovoltaic solar panels and hybrid solar panels. The heat pump system – Commotherm Hybrid Tower WW, Split DeLuxe (Austria) with a heating capacity of 5.7 kW is equipped with a two-section storage tank, the lower section of which has a volume of 200 litres, used as a low-potential energy source connected to hybrid solar collectors. The upper section of it has a volume of 300 litres, used as a heat accumulator connected to hybrid solar collectors. The following levels of operation of the heat pump system have been established for the change in the temperature of the refrigerant at the inlet to the condenser and at the outlet from the condenser: first level – 45–40.5°C; second level – 50–45.5°C; third level – 55–50.5°C; fourth level – 60–55.5°C. They correspond to changes in the local water temperature: 35–40°C; 40–45°C; 45–50°C; 50–55°C. The temperature of the low-potential energy source – ethylene glycol solution at the outlet of the heat pump evaporator is: 16.3°C, 12.3°C, 8.6°C, 4.9°C, corresponding to the established operating levels for heating up to a temperature of 20°C. The following levels of operation of the hot water supply have been established for the change in the temperature of the water from hybrid solar collectors at the inlet to the heat accumulator and at the outlet from the heat accumulator: first level: 60–57.5°C; second level: 65–61°C; third level: 70–65°C; fourth level: 75–69°C. They correspond to changes in the local water temperature: 30–55°C. Warming heat carrier consumption – 0.095 kg/s. One of the most problematic areas is the harmonization of energy production and consumption in the context of distributed energy generation using renewable sources. Connecting to Smart Grid technologies will prevent the peak load of the power system under conditions of regulated voltage and a connection to the heat pump power supply and hot water power supply. According to formulas (1)–(3), the results of complex mathematical modelling of the heat pump power supply and hot water supply using hybrid solar collectors are presented (Tables 7.1–7.6).

Table 7.1. Operating parameters of the heat pump system

Levels of operation	G_r, kg/s	N_e, kW	N_t, kW	U, V	f, Hz	n, rpm	COP
first level	0.0340	0.705	1.4	194.2	24.28	738.4	8.08
second level	0.0350	0.923	2.9	254.3	31.78	953.4	6.17
third level	0.0357	1.185	4.3	326.5	40.81	1224.3	4.80
fourth level	0.0363	1.452	5.7	400	50	1500	3.92

Note: G_r – refrigerant consumption, kg/s; N_e – power of the compressor electric motor, kW; N_t – thermal power of the low-grade energy source, kW; U – voltage, volts; f – current frequency, Hz; n – the number of revolutions of the compressor electric motor, rpm; COP – coefficient of performance of the heat pump system.

Table 7.2. Operating parameters of the hot water supply

Levels of operation	Gw, kg/s	Nt, kW	U, V	f, Hz	n, rpm
first level	0.0090	1	92	20	1140
second level	0.0100	1.5	138	30	1710
third level	0.0106	2	184	40	2280
fourth level	0.0120	2.5	230	50	2850

Note: Gw – local water consumption, kg/s; Nt – heating power, kW; U – voltage, volts; f – voltage frequency, Hz; n – the number of revolutions of the circulation pump electric motor, rpm.

Table 7.3. Heat transfer parameters as part of complex mathematical modelling of heat pump power supply

Levels of operation	Parameters		
	α_r, kW /(m^2·K)	α_w, kW /(m^2·K)	k, kW /(m^2·K)
first level	1.15	0.785	0.460
second level	1.19	0.901	0.505
third level	1.26	1.076	0.570
fourth level	1.38	1.372	0.674

Note: αr – coefficient of convective heat transfer from the refrigerant to the condenser wall, kW /(m2·K); αw – coefficient of convective heat transfer from the condenser wall to local water, kW /(m2·K); k – heat transfer coefficient, kW /(m2·K).

Table 7.4. Heat transfer parameters as part of complex mathematical modelling of the hot water supply

Levels of operation	Parameters		
	α_c, kW /(m^2·K)	α_w, kW /(m^2·K)	k, kW /(m^2·K)
first level	1.513	0.749	0.486
second level	1.533	0.851	0.530
third level	1.549	0.927	0.560
fourth level	1.555	0.953	0.564

Note: αc – coefficient of convective heat transfer from the warming heat carrier to the heat exchanger wall, kW /(m2·K); αw – coefficient of convective heat transfer from the heat exchanger wall to local water, kW /(m2·K); k – heat transfer coefficient, kW /(m2·K).

Table 7.5. Time constants and coefficients of the mathematical model of the dynamics of the temperature of the water of the heat pump power supply

Levels of operation	T_w, s	T_m, s	L_w, m	ε	L_r, m	L_r^*	ε^*	ζ
first level	6.23	2.62	24.61	1.7756	2.50	0.2857	1.2683	0.6811
second level	5.42	2.28	21.43	1.6014	2.48	0.2874	1.1412	0.6309
third level	4.54	1.91	17.95	1.4182	2.44	0.2907	1.0059	0.5945
fourth level	3.56	1.50	14.08	1.2204	2.30	0.3030	0.8506	0.5794

Note: Tw and Tm – time constants characterizing the local water thermal storage capacity, metal, seconds; Lw, Lr, Lr*, ε, ε*, ζ – coefficients of the mathematical model of the local water temperature dynamics.

Table 7.6. Time constants and coefficients of the mathematical model of the dynamics of the temperature of the water of the hot water power supply

Levels of operation	Tw, s	Tm, s	Lw., m	ε	Lc., м	Lc*	ε*	ζ
First level	37.4	7.96	0.59	1.72	3.65	0.22	1.34	2.8
Second level	32.9	7.0	0.58	1.53	3.61	0.22	1.19	2.85
Third level	30.2	6.43	0.56	1.42	3.57	0.22	1.11	2.95
Fourth level	29.4	6.26	0.62	1.39	3.55	0.22	1.08	2.66

Note: Tw, Tm – time constants characterizing the local water thermal storage capacity, metal, seconds; Lw, Lc, Lc*, ε, ε*, ζ – coefficients of the mathematical model of the local water temperature dynamics.

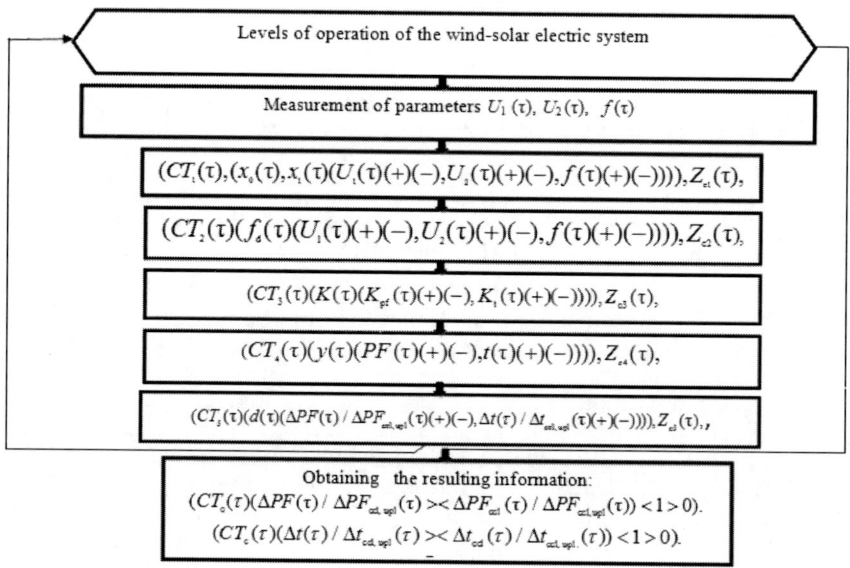

Figure 7.3. Block diagram of the wind-solar electric system operational control: U_1, U_2 – voltage at the network inverter input and the frequency converter output; f – current frequency, Hz; PF – power factor of the wind-solar electric system; t – local water temperature, °C; CT – event control; Z – logical relations; d – dynamic parameters; x – effects; f_d – parameters measured; y – parameters predicted; K – coefficients of mathematical description; ι – time. Indices: c – control of operability; ccl upl – the constant calculated value of the parameter of the lower and upper levels of operation (heat pump power supply), (hot water power supply), respectively; ccl – the constant calculated value of the operation level parameter; 0,1,2 – initial stationary mode, external, internal influences; 3 – coefficients of dynamics equations; 4 – significant predicted parameters; 5 – dynamic parameters.

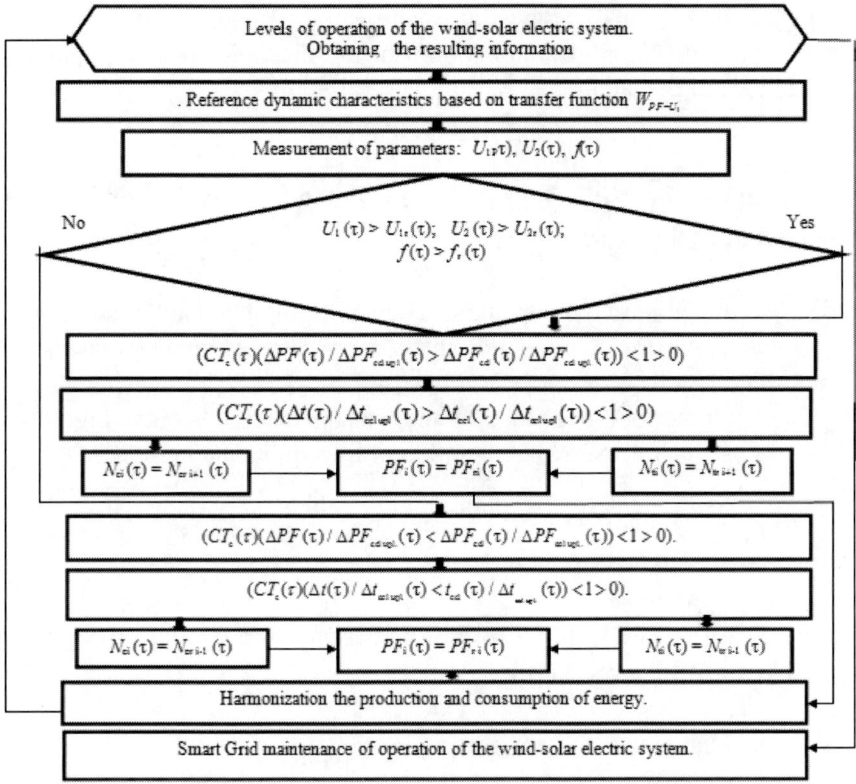

Figure 7.4. Block diagram of operational maintenance of the wind-solar electric system: U_1, U_2 – voltage at the network inverter input and the frequency converter output; f – current frequency, Hz; PF – power factor of the wind-solar electric system; N_e, N_t – power of the compressor electric motor, heating power, respectively, kW; ι – time; Indices: i – number of operation levels; r – reference value of the parameter; ccl upl – constant calculated value of the parameter of the lower and upper levels of operation (heat pump power supply), (hot water power supply), respectively; ccl – constant calculated value of the operation level parameter.

As presented in Tables 7.5 and 7.6, the time constants and coefficients that are part of the mathematical model of dynamics (3) are obtained based on parameters as part of complex mathematical modelling (Tables 7.1–7.4). Based on the proposed mathematical substantiation for operational maintenance of the Smart Grid wind-solar electric system (1) to (4), the block diagram for the operability control of the wind-solar electric system (Figure 7.3) is developed.

Operability control of the wind-solar electric system (Figure 7.3) enables obtaining the resulting information on decision-making about the operational maintenance of the wind-solar electric system.

Maintaining the Voltage within the Distribution System Based on Coordination of Energy Production and Consumption

Based on the proposed mathematical substantiation (1) to (4), the block diagram for operational maintenance of the Smart Grid wind-solar electric system (Figure 7.4) is developed.

The coordination of production and consumption of energy (Figure 7.4) makes it possible to ensure the operation of the wind-solar system.

Results and Discussion

Harmonization of Electric Power Production and Consumption in the Heat Pump Power Supply

A complex integrated system for supporting the operation of heat pump power supply has been developed (Table 7.7, Figures 7.5, 7.6). This system is based on the predicted power factor and local water temperature changes. The voltage at the grid inverter input, the voltage at the frequency converter output and the current frequency are continuously measured. Advance decisions are made to change the power of the heat pump compressor electric motor in accordance with the change in the thermal power of the lower section of the two-section storage tank as a low potential energy source.

The power factor of the wind-solar electric system in the established period is determined as follows:

$$PF_{i+1}(\tau) = PF_i + \\ + \begin{pmatrix} \Delta PF_{i+1}(\tau) / \Delta PF_{1..}(\tau) - \\ -\Delta PF_i(\tau) / \Delta PF_1(\tau) \end{pmatrix} (PF_2 - PF_1), \tag{17}$$

where PF is the power factor of the wind-solar electric system; PF_1 and PF_2 are the initial, final values of power factor; τ is the time, seconds. Index: 1 is the constant, calculated value of the parameter of the lower operational level; i is the number of levels of the wind-solar electric system operation.

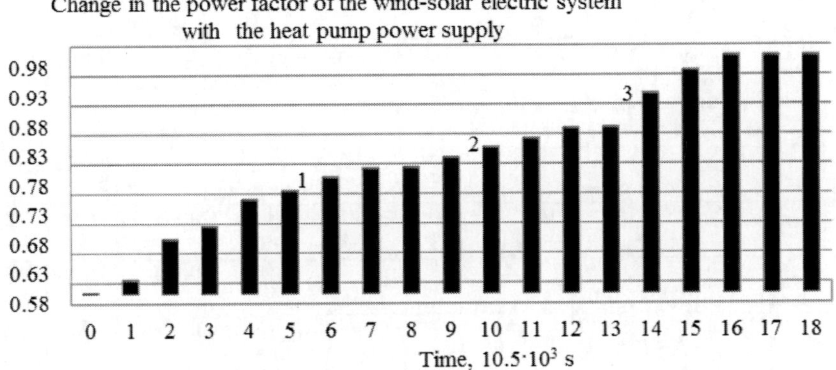

Figure 7.5. Operational maintenance of the wind-solar electric system. 1, 2, and 3 are the points when decisions were made on changing the power level in the electric motor of the heat pump compressor in accordance with the change in the thermal power of the lower section of the two-section storage tank as a low-potential energy source.

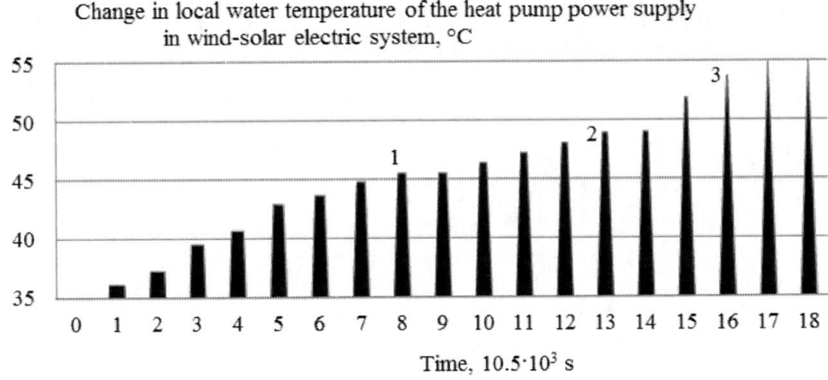

Figure 7.6. Operational maintenance of the local water temperature of the heat pump power supply in the wind-solar electric system. 1, 2, and 3 are the points when decisions were made on changing the power level in the electric motor of the heat pump compressor in accordance with the change in the thermal power of the lower section of the two-section storage tank as a low-potential energy source.

Table 7.7. Integrated Smart Grid system for harmonization of electric power production and consumption in the heat pump power supply

Time, τ, 103 s	Changing the parameters of the technological process	ΔPF(τ) /ΔPF1(τ)	PF(τ)	tw(τ), °C
0	Charging U1 = 84 V; U2 = 190.7 V; f = 25.09 Hz; Gr. = 0.0340 kg/s; tr in = 45°C; tr out. = 40.5°C; tw in = 35°C; t w out. = 40°C	1	0.58	35
10.5	U1 = 90 V U2 = 190.7 V; tr in = 45°C; tr out = 40.5°C; tw in = 35°C; t w out. = 40°C	0.9438	0.6025	36.12
21	U1 = 96 V; U2 = 190.7 V; tr in = 45°C; tr out = 40.5°C; tw in = 35°C; t w out. = 40°C	0.8875	0.6250	37.25
31.5	U1 = 108 V; U2 = 190.7 V; tr in = 45°C; tr out = 40.5°C; tw in = 35°C; t w out. = 40°C	0.7751	0.6700	39.5
42	U1 = 114 V; U2 = 190.7 V; tr in = 45°C; tr out = 40.5°C; tw in = 35°C; t w out. = 40°C	0.7188	0.6925	40.63
52.5	U1 = 126 V; U2 = 190.7 V; tr in = 45°C; tr out = 40.5°C; tw in = 35°C; t w out. = 40°C	0.6064	0.7375	42.88
63	U1 = 130 V; U2 = 190.7 V; tr in = 45°C; tr out = 40.5°C; tw in = 35°C; t w out. = 40°C	0.5689	0.7525	43.63
73.5	U1 = 136 V; U2 = 190.7 V; tr in = 45°C; tr out = 40.5°C; tw in = 35°C; t w out. = 40°C	0.5126	0.7750	44.76
84	U1 = 140 V; U2 = 190,7 V; tr in = 45°C; tr out = 40.5°C; tw in = 35°C; t w out. = 40°C	0.4752	0.7900	45.51
94.5	Decision on discharging Gr = 0.035 kg/s; U1 = 173.96 V; U2 = 241.56 V; f = 31.78 Hz; tr in = 50°C; tr out. = 45.5°C; tw in = 40°C; t w out. = 45°C	0.4783	0.7912	45.51
105	Charging U1 = 179.96 V; U2 = 241.56 V; f = 31.78 Hz; tr in = 50°C; tr out. = 45.5°C; tw in = 40°C; tw out. = 45°C	0.4355	0.8083	46.37
115.5	U1 = 185.96 V; U2 = 241.56 V; tr in = 50°C; tr out. = 45,5°C; tw in = 40°C; tw out. = 45°C	0.3931	0.8253	47.22
126	U1 = 191.96 V; U2 = 241.56 V tr in = 50°C; tr out. = 45.5°C; tw in = 40°C; tw out. = 45°C	0.3506	0.84	48.07
136.5	U1 = 197.96 V; U2 = 241.56 V tr in = 50°C; tr out. = 45,5°C; tw in = 40°C; tw out. = 45°C	0.3082	0.8570	48.92
147	Decision on discharging Gr = 0.0357 kg/s; U1 = 257.9 V; U2 = 310.17 V; f = 40.81 Hz; tr in = 55°C; tr out. = 50.5°C; tw in = 45°C; tw out. = 50°C	0.3125	0.8586	49
157.5	Charging U1 = 281.9 V; U2 = 310.17 V; f = 40.81 Hz; tr in = 55°C; tr out. = 50.5°C; tw in = 45°C; tw out. = 50°C	0.1690	0.9154	51.87
161.6	Decision on discharging Gr = 0.0363 kg/s; U1 = 341.93 V; U2 = 380 V; f = 50 Hz; tr in = 60°C; t r out. = 55.5°C; tw in = 50°C; tw out. = 55°C	0.0690	0.9554	53.67

Time, τ, 103 s	Changing the parameters of the technological process	ΔPF(τ) /ΔPF1(τ)	PF(τ)	tw(τ), °C
161.6	Discharging U1 = 341.93 V; U2 = 380 V; f = 50 Hz; Gr = 0.0363 kg/s; tr in = 60°C; t r out. = 55.5°C; tw in = 50°C; tw out. = 55°C	0	0.98	55

Note: PF – power factor; tr in, t r out, tw in, tw out – refrigerant temperature, local water temperature at the inlet to the condenser, at the outlet from the condenser, °C; Gr – refrigerant consumption, kg/s; f – current frequency, Hz; U1, U2 – voltage at the network inverter input and the frequency converter output, volts; τ – time, seconds. Indices: w – internal flow – local water; 1 – constant, calculated value of the parameter of the lower operational level.

The local water temperature in the established period is determined as follows:

$$t_{wi+1}(\tau)=t_{wi}+\left(\left(\frac{\Delta t_{wi+1}(\tau)}{\Delta t_{1.}(\tau)}-\frac{\Delta t_{wi}(\tau)}{\Delta t_{1}(\tau)}\right)(t_{w2}-t_{w1})\right), \tag{18}$$

where t_w – local water temperature, °C; t_1, t_2 – initial and final values of local water temperature, °C, respectively; i – the number of operational levels; τ – time, seconds. Index 1 – the constant, calculated value of the lower operational level parameter.

So, for example, in the period of time $147\cdot10^5$ s (4083 h) from the beginning of the heating season, the predicted increase in the power factor went from 0.8570 to 0.8586. The power factor using formula (17) is:

0.8586 = 0.8570 + (0.3125-0.3080)(0.98-0.95)

In the period of time $147\cdot10^5$ s (4083 h), the predicted increase in the local water temperature went from 48.92°C to 49°C. The local water temperature using formula (18) is:

49°C = 48.92°C+(0.3125–0.3080)(55°C–35°C).

During this period of time, when the voltage at the grid inverter input changes to a level of 257.9 V, it is necessary to make a decision to maintain the power factor of the wind-solar electrical system at the level of 0.8586. Predicting an increase in the electric grid charge, it is necessary to increase the power of the heat pump compressor electric motor based on the current frequency change to 40.81 Hz. Adopting a proactive decision to increase the number of revolutions of the compressor electric motor enables increasing the refrigerant consumption to the level of 0.0357 kg/s and ensuring the maintenance of the local water temperature at 49°C. The implementation of such actions will enable coordinating the production and consumption of energy while maintaining the operation of the heat pump power supply.

Harmonization of Electric Power Production and Consumption in the Hot Water Power Supply

A complex integrated system for supporting the operation of the hot water power supply has been developed (Table 7.8, Figures 7.7, 7.8). This system is based on the predicted power factor and local water temperature changes. The voltage at the grid inverter input, the voltage at the frequency converter output and the current frequency are continuously measured. Advance decisions are made to change the power of the electric motor of the circulation pump, based on the change in the current frequency in accordance with the change in the thermal power of the upper section of the two-section storage tank.

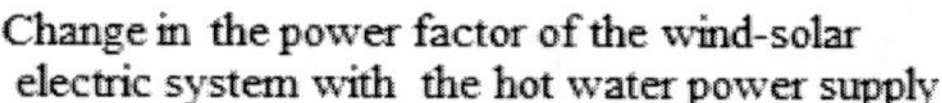

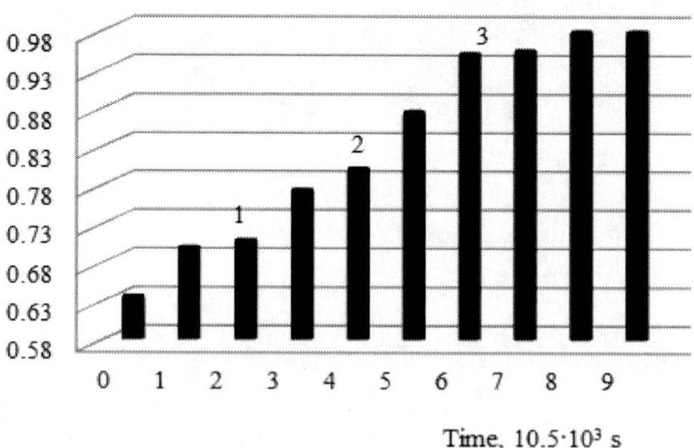

Figure 7.7. Operational maintenance of the wind-solar electric system. 1, 2, and 3 are the points when decisions were made on changing the power level in the electric motor of the circulation pump in accordance with the change in the thermal power of the upper section of the two-section storage tank.

Table 7.8. Integrated Smart Grid system for harmonization of electric power production and consumption in the hot water power supply

Time, τ, 103 s	Changing the parameters of the technological process	ΔPF(τ) /ΔPF1(τ)	PF(τ)	tw(τ), °C
3	Charging U1 = 32 V; U2 = 38.8 V; f = 8.43 Hz; Gw = 0.009 kg/s; tin = 60°C; tout = 57.5°C	0.1435	0.6374	33.59
6	U1 = 72 V; U2 = 86.27 V; f = 18.75 Hz; Gw = 0.009 kg/s; t in = 60°C; tout = 57.5°C	0.3010	0.7004	37.53
9	Decision on discharging Gw = 0.0100 kg/s; U1 = 72 V; U2 = 86.27 V; f = 18.75 Hz; tin = 65°C; tout = 61°C	0.3256	0.7102	38.14
12	Charging U1 = 108 V; U2 = 129.4 V; f = 28.13 Hz; Gw = 0.010 kg/s; tin = 65°C; tout = 61°C	0.4883	0.7753	42.21
15	Decision on discharging Gw = 0.0106 kg/s; U1 = 108 V; U2 = 129.4 V; f = 28.13 Hz; tin = 70°C; tout = 65°C	0.5567	0.8027	43.92
18	Charging U1 = 144 V; U2 = 172.5 V; f = 37.5 Hz; Gw = 0.0106 kg/s; tin = 70°C; tout = 65°C	0.7415	0.8766	48.54
21	U1 = 180 V; U2 = 215.7 V; f = 46.9 Hz; Gw = 0.0106 kg/s; tin = 70°C; tout = 65°C	0.9288	0.9515	53.22
24	Decision on discharging Gw = 0.0120 kg/s; U1 = 180 V; U2 = 215.7 V; f = 46.9 Hz; tin = 75°C; tout = 69°C	0.9395	0.9558	53.49
27	Discharging U1 = 182 B; U2 = 230 B; f = 50 Hz; Gw = 0.0120 kg/s; tin = 75°C; tout = 69°C	1	0.98	55

Note: PF – power factor; tin, tout – warming heat carrier temperature at the inlet to the heat exchanger, at the outlet from the heat exchanger, °C; tw – local water temperature, °C; Gw – local water consumption, kg/s; f – current frequency, Hz; U1, U2 – voltage at the network inverter input and the frequency converter output, volts; τ – time, seconds. Indices: w – local water; 1 – constant, calculated value of the parameter of the upper operational level.

Thus, for example, in the period of time $15 \cdot 10^3$ s (4.2 h) from the beginning of the predicted increase power factor of the solar radiation from 0.7753 to 0.8027. The power factor using formula (17) is:

$$0.8027 = 0.7753 + (0.5567\text{-}0.4883)(0.98\text{-}0.95).$$

In this period of time $15 \cdot 10^3$ s (4.2 h), the predicted increase in the local water temperature went from 42.21°C to 43.92°C. The local water temperature using formula (18) is:

$$43.92°C = 42.21°C + (0.5567\text{-}0.4883)(55°C–30°C).$$

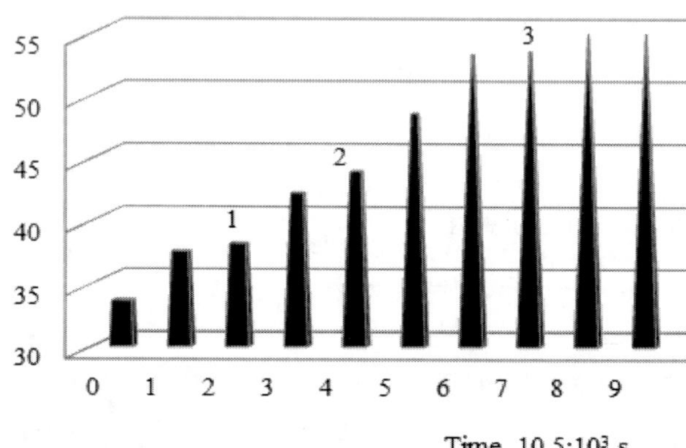

Figure 7.8. Operational maintenance of the local water temperature of the hot water supply in the wind-solar electric system. 1, 2, and 3 are the points when decisions were made on changing the power level in the electric motor of the circulation pump in accordance with the change in the thermal power of the upper section of the two-section storage tank.

During this period of time, when the voltage at the grid inverter input changes to a level of 108 V, it is necessary to make a decision to maintain the power factor of the wind-solar electrical system at the level of 0.8027. Predicting an increase in the electric grid charge, it is necessary to increase the power of the circulating pump electric motor based on the current frequency change to 28.13 Hz. Adopting a proactive decision to increase the number of revolutions of the electric motor enables increasing the local water consumption to the level of 0.0106 kg/s and ensuring the maintenance of the local water temperature at 43.92°C. The implementation of such actions will enable the coordination of the production and consumption of energy while maintaining the operation of the hot water power supply.

Conclusion

This book proposes a mathematical substantiation of operational maintenance of Smart Grid wind and solar electric systems based on the architecture and mathematical description of the architecture of wind and solar electric systems using the methodology for the mathematical description of the dynamics of power systems, and the cause-effect graph method. The mathematical substantiation is based on a systematic approach to complex mathematical and logical modelling in the wind and solar electric systems as dynamic systems. Decision-making to support the operation of energy systems is based on a mathematical description of the architecture of wind and solar electric systems, the methodology of the mathematical description of the dynamics of energy systems, and the cause-effect graph method. Coordination of energy production and consumption is based on forecasting changes within the parameters of technological processes. Real wind and solar electric systems are dynamic systems, the mathematical models of which reflect the properties of the transformation of influences, that is, their dynamic properties. The operational maintenance of wind and solar electric systems takes place in the composition of such wind and solar electric systems, which are based on dynamic systems. When designing wind and solar systems, integrated dynamic subsystems for evaluating changes in both energy production and consumption are laid at their foundations. By representing the design of technological systems as an organization of complex systems, it is expanded by building dynamic subsystems on its foundation – nodes that predict the components of technological processes. Based on the structured system and mathematical substantiation of the architecture of the wind and solar electric systems, we consider the relation category as organizing interactions not only within the elements of the technological systems but also within the elements of the integrated dynamic subsystems, which makes it possible to perform operability control and identify the state of the energy systems based on the developed cause-effect graph method. The final assessment, obtained by using logical relations in the dynamic subsystems, enables the use of new properties of the energy systems as a result of appropriate decision-making and identification of new operating conditions. Moreover, the relations between the dynamic subsystems and the blocks in their composition enable, based on the assessment of the state of the parameters diagnosed in these blocks, to confirm the new conditions of the energy systems functioning.

Based on mathematical and logical modelling, a technological support system was developed for changing the battery capacity created on the predicted voltage variation by measuring the temperature of electrolyte within the battery volume as a part of a technological system for battery operation. The developed technology makes it possible: to control the operational capacity of the accumulator battery in order to obtain a functional assessment of change in the total charge and discharge voltage; to obtain an integrated reference estimation of change in the charge and discharge voltage; to develop an integrated system for assessing a change in the voltage of the battery, which enables maintaining capacity of the accumulator battery by measuring the temperature of electrolyte at the input to the battery. The maximum electrolyte temperature change – 35°C, was set when charging from a direct current supply, and a limit in voltage change for a further charge and discharge was established with a change in the consumption of electric energy. The use of an integrated system for the estimation of voltage change obtained based on the alignment between electrochemical and diffusion processes of discharge and charge enables making timely decisions on recharging to prevent overcharging and unacceptable discharge. Coordination of the electrochemical and diffusion processes that accompany the charging and discharging of the battery makes it possible, for example when operating a wind power plant with a capacity of 10 kW, to reduce the cost of energy production and the payback period of the wind power plant by up to 25% due to a reduction of the charge period and prevention of gas formation.

By utilizing mathematical and logical modelling as part of the technological system for the operation of thermal electric accumulators, a system of technological support for changing the capacity of accumulators was developed based on predicting changes in the temperature of heated water within the accumulator volume when the water temperature is measured at the thermoelectric accumulator inlet and outlet. The technology developed enables controlling the operation of a thermal electric accumulator in order to obtain a functional assessment of the change in the temperature of the heated water. The use of an integrated system for estimating the change in the temperature of heated water within the battery volume, obtained by matching the heat and mass transfer processes during discharging and charging, makes it possible to make timely decisions on changing the power of a thermal electric accumulator based on a change in the local water flow rate, which enables reducing the charging time by up to 30%. For example, this technology is used in the developed integrated Smart Grid system to support the operation of a wind-solar power plant by predicting changes in the battery capacity (Chapter 6). Changes in the speed of rotation of the circulation pump electric motor are provided under the conditions of changes in the flow rate and temperature of the heated water by reducing the charging duration by up to 30%. The storage battery and the thermoelectric accumulator as part of the network solar power system (Chapter 4) acquire the additional status of voltage regulators within the distribution system. Preliminary decisions are made to change the capacitance of a thermoelectric accumulator when redistributing the accumulated electrical energy. Changing the level of transmission of electrical energy to the grid enables maintaining the voltage within the distribution system by maintaining the power factor of the solar power plant grid. The peak load on the power system is prevented, which reduces the consumption of electricity from the network by up to 14%.

It is proposed to maintain the voltage within the distribution system by forecasting the change in the storage battery capacity in order to make anticipatory decisions on the change in the power of the thermoelectric accumulator for the purpose of the maintenance of the power factor of the solar electric system network. The change in the ratio of the voltage at the frequency converter output and the voltage within the distribution system is assessed by measuring the voltage at the input of the hybrid inverter. A comprehensive system has been developed to support the operation of a solar power plant network based on predicting changes in battery capacity and power factor. The accumulator battery and the thermoelectric accumulator as part of the solar electric system network acquire the additional status of voltage regulators in the distribution system. Preliminary decisions are made to change the capacity of a thermoelectric accumulator to redistribute the accumulated electrical energy. Changing the level of transmission of electrical energy to the grid enables maintaining the voltage in the distribution system by maintaining the power factor of the solar power plant grid. Continuous measurement of the voltage at the hybrid inverter input, at the frequency converter output, and within the distribution network is performed. The change in the ratio of voltages at the frequency converter output and within the distribution network is estimated. Peak load on the power system is prevented, reducing power consumption from the grid by up to 14%.

The Integrated Smart Grid System of electric power production and consumption harmonization based on a prediction of changes in the battery capacity is developed. The integrated dynamic subsystem of the photoelectric charging station includes the following components: mains, photoelectric solar panels, a hybrid inverter, rechargeable batteries, a two-way Smart Meter and a charger. Advance decisions on the change in power transmission capacity enabled regulating the voltage within the distribution system by maintaining the power factor of the photoelectric charging station. Voltages at the hybrid inverter input and within the distribution system were measured to assess their ratio. Comprehensive mathematical and logical modelling of the photoelectric charging station was performed based on the mathematical substantiation of architecture and operational maintenance. A dynamic subsystem including such components as mains, a photoelectric module, a hybrid inverter, batteries, a two-way Smart Meter and a charger formed the basis of the proposed technological system. Time constants and coefficients of dynamic mathematical models were determined in terms of the estimation of changes in the battery capacity and power factor of the photoelectric charging station. A functional estimate of changes in the battery capacity and power factor of the photoelectric charging station was obtained. Maintenance of voltage within the distribution system was realized based on the resulting operational data to estimate a change in the battery capacity. Advance decision-making has made it possible to raise the power factor of the photoelectric charging station by up to 40% due to matching the electric power production and consumption. Operational maintenance of the photoelectric charging station using the developed Smart Grid technology has enabled the prevention of peak loading of the power system due to a 20% reduction in power consumption from the network.

The developed integrated Smart Grid system for supporting the operation of a wind-solar electric system is based on predicting a change in the capacity of a rechargeable battery by measuring the voltage at the hybrid charge controller input, the voltage at the inverter output, and the current frequency. Making advance decisions to support the capacity of a rechargeable battery related to a change in the capacity of a thermoelectric battery is based on establishing the ratio of voltage measured at the hybrid charge controller input to voltage at the inverter output. A change in the rotational speed of the electric motor of the circulating pump has been ensured in terms of changes in consumption and the temperature of heated water by reducing charge duration by up to 30%. The basis for the proposed technological system is a dynamic subsystem that includes the following components: a wind-energy installation, a photoelectrical electrical module, a hybrid charge controller, an inverter, an array of rechargeable batteries, and a thermoelectric battery. A functional assessment has been derived for a change in the capacity of a rechargeable battery, the rotational speed of the electric motor of the circulating pump, and local water consumption related to a change in the local water temperature in the range of 30–70°C. Defining the resulting functional information on forecasting a change in the capacity of a rechargeable battery enables making the following advance decisions on changing the rotational speed of the electric motor of the circulating pump and the consumption of local water. Maintaining the capacity of a rechargeable battery is carried out based on adjusting the generation and consumption of energy.

Integrated Smart Grid Systems for harmonization of production and consumption of electric power of the heat pump power supply and hot water power supply are being developed. The integrated dynamic subsystem of the wind-solar electric system includes the following components: the electric network, wind turbine, photovoltaic solar panels, hybrid solar collectors, grid inverter, heat pump, two-section storage tank, upper section for hot water supply, lower section a low-grade energy source, frequency converter. Integrated systems based on predicting changes in the power factor and local water temperature by measuring voltage from hybrid solar collectors at the grid inverter input, voltage at the frequency converter output, as well as current frequency. The adoption of advance decisions to maintain the local water temperature by changing the power of the electric motor of the heat pump compressors and the electric motor of the circulation pump is based on establishing the ratio of the voltages at the grid inverter input and the frequency converter output. The power factor of the wind-solar electric system is maintained. The complex mathematical and logical modelling of the wind-solar electric system, based on the mathematical substantiation of the architecture of the wind-solar electric system and mathematical substantiation of the operational maintenance of the wind-solar electric system, is performed. Time constants and coefficients of the mathematical models of dynamics regarding the estimation of a change in the power factor of the system, and the local water temperature are predicted by measuring voltage from hybrid solar collectors at the grid inverter input, voltage at the frequency converter output and current frequency. Functional estimation of a change in the power factor of the wind-solar electric system is in the range of 58–98%, and local water temperature is in the range of 30–55°C. Determining final functional information provides an opportunity to make advance decisions on a change in the operation of the electric motor in the heat pump compressor and the electric motor in the circulation pump to prevent the peak load on the power system and maintain voltage when the heat pump power supply and hot water supply are connected.

References

Chaikovskaya E. (2022). Book. *Smart Grid Technologies in Electric Systems for Renewable Energy. Energy Science, Engineering and Technology*. ISBN 979-8-88697-387-7 Published by NOVA Science Publishers Inc. New York. P.181.

Chaikovskaya E. (2021). Book. Chapter 6. Smart Grid Technology for Maintaining the Functioning of a Wind-Solar Electric System. *Advances in Energy Research. Volume 35.* Morena J. Acosta (Editor). NOVA Science publishers. New York. P.215-236.

Chaikovskaya, E. (2014). Maintaining the relation between production and consumption of electricity and heat at decision-making level. *Eastern-European Journal of Enterprise Technologies* (3(8)), 4-9.

Chaikovskaya, E. (2015). Devising an energy saving technology for a biogas plant as a part of the cogeneration system. *Eastern-European Journal of Enterprise Technologies*, (3/8(75)), 47-53. doi: https://doi.org/10.15587/1729-4061.2015.44252.

Chaikovskaya, E. (2016). The development of energy-saving operation technology of the biodiesel plant as a part of the cogeneration system. *Eastern-European Journal of Enterprise Technologies*, (1/8(79)), 4-11. doi: https://doi.org/10.15587/1729-4061.2016.59479.

Chaikovskaya, E. (2016). Development of energy-saving technology maintaining the functioning of a drying plant as a part of the cogeneration system. *Eastern-European Journal of Enterprise Technologies*, (3/8(81)), 42-46. doi: https://doi.org/10.15587/1729-4061.2016.72540.

Chaykovskaya, E. E. (2016). Soglasovanie proizvodstva i potrebleniya energii na osnove intellektual'nogo upravleniya teplomassobmennymi protsessami. XV Minskiy mezhdunarodnyy forum po teplomassobmenu: materialy XV Minsk. mezhd. foruma po teplomassobmenu. Sektsiya 8. Teplomassoperenos v energeticheskih protsessah i oborudovanii. *Energosberezhenie*. Minsk, 1-12.

Chaikovska, E. E. (2017). Intelectualni systemi pidtrymky functcionuvania energetichnih system na rivni priniattia rihen. *Enerhetyca. economica, tehnologii, ekologia*, 3, 114-118. doi: https://doi.org/10.20535/1813-5420.3.2017.117377.

Chaikovska, E. E. (2016). Informatsiyni tekhnolohiyi pidtrymky funktsionuvannia enerhetychnykh system na rivni pryiniattia rishen. *Informatyka. Kultura. Tekhnika: zb. tez dop. IV ukr.-nim. konf. Odesa*, 32-33.

Chaikovskaya, E. (2017). Development of energy-saving technology to support functioning of the lead-acid batteries. *Eastern-European Journal of Enterprise Technologies*, (4/8(88)), 56-64. doi: https://doi.org/10.15587/1729-4061.2017.108578.

Chaikovskaya, E. (2018). Development of energy-saving technology for maintaining the functioning of heat pump power supply. *Eastern-European Journal of Enterprise Technologies*, (4/8(94)), 13-23. doi: https://doi.org/10.15587/1729-4061.2018.139473.

Chaikovskaya, E. (2019). Development of energy-saving technology to maintain the functioning of a wind-solar electrical system. *Eastern-European Journal of Enterprise Technologies*, (4/8(100)), 57-68. doi: https://doi.org/10.15587/1729-4061.2019.174099.

Chaikovskaya, E. (2020). Development of Smart Grid technology for maintaining the functioning of a biogas cogeneration system. *Eastern-European Journal of Enterprise Technologies*. (8/3(105)), 56-68. doi: https://doi.org/10.15587/1729-4061.2020.205123.

Chaikovskaya, E. (2021). Development of Smart Grid technology to maintain the functioning of photoelectric charging stations. *Eastern-European Journal of Enterprise Technologies*. (3/8(111)), 14-24. doi: https://doi.org/10.15587/1729-4061.2021.235120.

Chaykovskaya, E. E. (2021). Kompleksnoie modelirovanie protcessov teplomassopierenosa kak osnova upravlienia na urovnie priniatia rechenii. XVI Minskiy mezhdunarodnyy forum po teplomassobmenu. *Tezisi dokladov I soobchenii XVI Minsk. mezhd. foruma po teplomassobmenu. Sektcia 8. Modelirovanie i upravlenie protcessami teplomassoperenosa*. 1290-1293.

Chaikovskaya, E. (2020). Complex mathematical modelling of heat pump power supply based on wind-solar network electrical system *Technology Audit and Production Reserves*. 6/1(56), 28-33. doi: https://doi.org/10.15587/2706-5448.2020.220269.

Chaikovska, E. E, Marenich V. E. (student) (2020). Matematichne obhruntuvannia arhitecturi kombinovanoi meregevoi fotoelectrichnoi sistemi. *Informatciini tehnologii v modeluvanni: Materiali V vseukrainskoi nauk.-prakt. konf. studentiv, aspirantiv ta molodih vchenih.* Odesa. ONPU. 97-100.

Chaikovska, E. E., Hega K. V. (student) (2020). Matematichne obhruntuvannia pidtrimki funktcionuvannia meregevoi vitro-soniachnoi electrichnoi sistemi. *Informatciini sistemi priiniattia richen i problemi obchisluvalnoho intelectu – ISDMCI'2020*: m*ateriali mignar. nauk. konf.* Kherson. 154-155.

Chaikovska, E. E., Hega K. V. (student) (2020). Matematichne obhruntuvannia teplonasosnoho enerhopostachannia u skladi meregevoi vitro-soniachnoi elektrichnoi sistemi. *Pratci 9 Mign. nauk.-prakt. konf. "Problemi informatiki ta komputernoi tehniri" (PIKT -2020),* Chernivtsi, 123-125.

Chaikovska, E. E., Shtapenko S. V. (student) (2021). Matematichne obhruntuvannia pidtrimki funktcionuvannia fotoelectrichnoi zariadnoi stantcii. *Tezi dop. XVII Mignarodn. nauk.-prakt. konf. «Enerhetichni ta teplotechnichni protcessi i ustatcuvannia»,* Kharkiv: HTU«HPI». 90–91. ISBN 978-617-7476-56-5.

Chaikovska, E. E. (2022). Kompleksne upravlinnia akumuluvanniam u skladi meregevoi soniachnoi elektrichnoi sistemi. *Pratci XI Mign. nauk.-prakt. konf. "Problemi informatiki ta komputernoi tehniri" (PIKT-2022),* Chernivtsi, 112-114.

Chaikovska, E. E. (2004). *Dopusk yak struktura.* Visnik Natcionalnoho universitetu “Lvivska politehnika.” Teploenergetika. Ingeneria dovkilla. Avtomatizatsia. Lviv. 506. 285-290.

Chaikovskaya, E. E. (1999) Sinergeticheski podhod pri razrabotke ekspertnih sistem. *Tr. Odes. Politehn.un-ta.* №2(8). 126-128.

Chaikovskaya, E. E (1999). Dinamicheskaya podsistema kak osnova ekspertnich system. *Tr. Odes. Politehn.un-ta.* №3(9). 108-110.

Chaikovskaya, E. E. (2000). Matematicheskoe modelirovanie dinamiki energeticheskih system kak osnovi diagnostiki *Tr. Odes. Politehn.un-ta.* №3(12). 83-86.

Al-Dhaifallah, M., Nassef, A. M., Rezk, H., Nisar, K. S. (2018). Optimal parameter design of fractional order control based INC-MPPT for PV system. SolarEnergy, 159, 650-664. doi: https://doi.org/10.1016/j.solener.2017.11.040.

Chen W. H., Che, Y. B. (2014). Design of Lead-Acid Battery Management System. *Applied Mechanics and Materials*, 533, 331–334. doi: https://doi.org/10.4028/www.scientific.net/amm.533.331.

Chunhe Song, Yingying Sun, Guangjie Han, Joel J. P. C. Rodrigues (2021). Intrusion detection based on hybrid classifiers for smart grid. *Computers & Electrical Engineering*, 93, 107212. doi: https://doi.org/10.1016/j.compeleceng.2021.107212.

Davye, M., Daranith, & Ch., Dae-Hyun, Ch. (2020). Sensitivity analysis of volt-VAR optimization to data changes in distribution networks with distributed energy resources. *Applied Energy*, 261,114331.

Dixon, J., Bell, K. (2020). Electric vehicles: Battery capacity, charger power, access to charging and the impacts on distribution networks. *eTransportation*, 4, 100059. doi: https://doi.org/10.1016/j.etran.2020.100059.

Dost, P., Martin, M., Sourkounis, C. (2014). Impact of lead-acid based battery design variations on a model used for a battery management system. *MedPower 2014*. doi: https://doi.org/10.1049/cp.2014.1691.

Fathabadi, H. (2017). Novel standalone hybrid solar/wind/fuel cell power generation system for remote areas. *Solar Energy*, 146, 30–43. doi: https://doi.org/10.1016/j.solener.2017.01.071.

Fathabadi, H. (2020). Novel stand-alone, completely autonomous and renewable energy based charging station for charging plug-in hybrid electric vehicles (PHEVs). Applied Energy, 260, 114194. doi: https://doi.org/10.1016/j.apenergy.2019.114194.

Gong, Y. L., Li, H. Z., Li, M. Q., Zhan, W. D. (2014). Design of Lead-Acid Battery Intelligent Charging System. *Applied Mechanics and Materials*, 651-653, 1068-1073. doi: https://doi.org/10.4028/www.scientific.net/amm.651-653.1068.

Guliaev V. A. (1990). Diagnostika entrgeticheskih i elektronnich sistem. *Sb. nauchnich trudov*. Kyiv. Naukova dumka.

Haken H. (1996). Principles of Brain Functioning. *Springer Series in Synergetics*. Berlin: Springer. 67.

Heydar Chamandoust, Abozar Hashemi, Salah Bahramara. (2021). Energy management of a smart autonomous electrical grid with a hydrogen storage system *International Journal of Hydrogen Energy,* 46, 34, 17608–17626. doi: https://doi.org/10.1016/j.ijhydene.2021.02.174.

Nishant Jha, Deepak Prashar, Mamoon Rashid, Sachin Kumar Gupta, R. K. Saket (2021). Electricity load forecasting and feature extraction in smart grid using neural networks. *Computers & Electrical Engineering,* 96, 107479. doi: https://doi.org/10.1016/j.compeleceng.2021.107479.

Palacky, P., Baresova, K., Sobek, M., Havel, A. (2016). The control system of electrical energy accumulation. *2016 ELEKTRO.* doi: https://doi.org/10.1109/elektro.2016.7512094.

Parminder Singh, Mehedi Masud, M. Shamim Hossain, Avinash Kaur. (2021). Blockchain and homomorphic encryption-based privacy-preserving data aggregation model in smart grid. *Computers & Electrical Engineering*, 93, 107209. doi: https://doi.org/10.1016/j.compeleceng.2021.107209.

Perera, A., Vahid M., & Wickramasinghe, P., Scartezzini, J. (2019). Redefining energy system flexibility for distributed energy system design. *Applied Energy*, 253, 113572.

Prigozin I. (1960). Vvedenie v termodinamiku neobratimih prosessov M. *Izd-vo inostr. lit-ri.* 160.

Saad, A. A., Faddel, S., Mohammed, O. (2019). A secured distributed control system for future interconnected smart grids. *Applied Energy*, 243, 57–70. doi: https://doi.org/10.1016/j.apenergy.2019.03.185.

Shahriari, M., Blumsack, S. (2018). The capacity value of optimal wind and solar portfolios. Energy, 148, 992-1005. doi: https://doi.org/10.1016/j.energy.2017.12.121.

Swathika, R., Ram, R. K. G., Kalaichelvi, V., Karthikeyan, R. (2013). Application of fuzzy logic for charging control of lead-acid battery in stand-alone solar photovoltaic system. *2013 International Conference on Green Computing, Communication and Conservation of Energy (ICGCE*). doi: https://doi.org/10.1109/icgce.2013.6823464.

Von Bertalanffy L. *General System Theory, Foundations, Development, Applications*: rev.ed. (1976). George Braziller, New York.

Yhosvany Soler-Castillo, Julio César Rimada, Luis Hernández, Gema Martínez-Criado. (2021). Modelling of efficiency of the photovoltaic modules: Grid-connected plants to the Cuban national electrical system. *Solar Energy,* vol. 223, pp.150-157. doi: https://doi.org/10.1016/j.solener.2021.05.052.

Yinan, L., Wentao, Y., & Ping, H.,Chang, Ch., Xiaonan, W. (2019). Design and management of a distributed hybrid energy system through smart contract and blockchain. *Applied Energy*, 248, 390–405.

Zaher, G. K., Shaaban, M. F., Mokhtar, M., Zeineldin, H. H. (2021). Optimal operation of battery exchange stations for electric vehicles. Electric Power Systems Research, 192, 106935. doi: https://doi.org/10.1016/j.epsr.2020.106935.

Zhiyong Li, Gang Chen. (2022). Fixed-time consensus based distributed economic generation control in a smart grid. *International Journal of Electrical Power & Energy Systems*, 134, 107437. doi: https://doi.org/10.1016/j.ijepes.2021.107437.

About the Author

Eugene Chaikovskaya

Affiliation: PhD, Senior Researcher.

Education: Department of Theoretical, General and Nonconventional Power Engineering, Odessa Polytechnic National University, Ukraine.

Business Address: Shevchenko Avenue, 1, Odesa, Ukraine, 65044.

Research and Professional Experience:

- Number of publications in Ukrainian publications: 233.
- Number of publications in indexed foreign publications: 14.
- ORCID: http://orcid.org/0000-0002-5663-2707
- http://www.scopus.com/authid/detail.uri?authorId=57170828500.
- ID: 57170828500.
- Scopus index: h index: 3; 10 individual publications (2014-2022).
- https://www.researchgate.net/profile/Eugene-Chaikovskaya.

Publications from years (2014-2022):

1. Chaikovskaya E. (2022). Book. *Smart Grid Technologies in Electric Systems for Renewable Energy. Energy Science, Engineering and Technology.* ISBN 979-8-88697-387-7 Published by NOVA Science Publishers Inc. New York. P.181.
2. Chaikovskaya E. (2021). Book. Chapter 6. Smart Grid Technology for Maintaining the Functioning of a Wind-Solar Electric System. *Advances in Energy Research. Volume 35.* Morena J. Acosta (Editor). NOVA Science publishers. New York. P.215-236. Publication Date:

November 11, 2021 Pages: 226. ISBN: 978-1-68507-373-1 (e-book) ISSN 2157-1562. doi: https://doi.org/10.52305/EEEQ2450.

3. Chaikovskaya, E. (2014). Maintaining the relation between production and consumption of electricity and heat at decision-making level. *Eastern-European Journal of Enterprise Technologies* (3(8)), 4-9.
4. Chaikovskaya, E. (2015). Devising an energy saving technology for a biogas plant as a part of the cogeneration system. *Eastern-European Journal of Enterprise Technologies*, (3/8(75)), 47-53. doi: https://doi.org/10.15587/1729-4061.2015.44252.
5. Chaikovskaya, E. (2016). The development of energy-saving operation technology of the biodiesel plant as a part of the cogeneration system. *Eastern-European Journal of Enterprise Technologies*, (1/8(79)), 4-11. doi: https://doi.org/10.15587/1729-4061.2016.59479.
6. Chaikovskaya, E. (2016). Development of energy-saving technology maintaining the functioning of a drying plant as a part of the cogeneration system. *Eastern-European Journal of Enterprise Technologies*, (3/8(81)), 42-46. doi: https://doi.org/10.15587/1729-4061.2016.72540.
7. Chaykovskaya, E. E. (2016). Soglasovanie proizvodstva i potrebleniya energii na osnove intellektual'nogo upravleniya teplomassobmennymi protsessami. XV Minskiy mezhdunarodnyy forum po teplomassobmenu: materialy XV Minsk. mezhd. foruma po teplomassobmenu. Sektsiya 8. Teplomassoperenos v energeticheskih protsessah i oborudovanii. *Energosberezhenie*. Minsk, 1-12.
8. Chaikovska, E. E. (2017). Intelectualni systemi pidtrymky functcionuvania energetichnih system na rivni priniattia rihen. *Enerhetyca. economica, tehnologii, ekologia*, 3, 114-118. doi: https://doi.org/10.20535/1813-5420.3.2017.117377.
9. Chaikovska, E. E. (2016). Informatsiyni tekhnolohiyi pidtrymky funktsionuvannia enerhetychnykh system na rivni pryiniattia rishen. *Informatyka. Kultura. Tekhnika: zb. tez dop. IV ukr.-nim. konf. Odesa*, 32-33.
10. Chaikovskaya, E. (2017). Development of energy-saving technology to support functioning of the lead-acid batteries. *Eastern-European Journal of Enterprise Technologies*, (4/8(88)), 56-64. doi: https://doi.org/10.15587/1729-4061.2017.108578.
11. Chaikovskaya, E. (2018). Development of energy -saving technology for maintaining the functioning of heat pump power supply. *Eastern-*

European Journal of Enterprise Technologies, (4/8(94)), 13-23. doi: https://doi.org/10.15587/1729-4061.2018.139473.

12. Chaikovskaya, E. (2019). Development of energy-saving technology to maintain the functioning of a wind-solar electrical system. *Eastern-European Journal of Enterprise Technologies*, (4/8(100)), 57-68. doi: https://doi.org/10.15587/1729-4061.2019.174099.
13. Chaikovskaya, E. (2020). Development of Smart Grid technology for maintaining the functioning of a biogas cogeneration system. *Eastern-European Journal of Enterprise Technologies*. (8/3(105)), 56-68. doi: https://doi.org/10.15587/1729-4061.2020.205123.
14. Chaikovskaya, E. (2020). Complex mathematical modelling of heat pump power supply based on wind-solar network electrical system *Technology Audit and Production Reserves*. 6/1(56), 28-33. doi: https://doi.org/10.15587/2706-5448.2020.220269.
15. Chaikovskaya, E. (2021). Development of Smart Grid technology to maintain the functioning of photoelectric charging stations. *Eastern-European Journal of Enterprise Technologies*. (3/8(111)), 14-24. doi: https://doi.org/10.15587/1729-4061.2021.235120.
16. Chaykovskaya, E. E. (2021). Kompleksnoie modelirovanie protcessov teplomassopierenosa kak osnova upravlienia na urovnie priniatia rechenii. XVI Minskiy mezhdunarodnyy forum po teplomassobmenu. *Tezisi dokladov I soobchenii XVI Minsk. mezhd. foruma po teplomassobmenu. Sektcia 8. Modelirovanie i upravlenie protcessami teplomassoperenosa*. 1290-1293.
17. Chaikovska, E. E, Marenich V. E. (student) (2020). Matematichne obhruntuvannia arhitecturi kombinovanoi meregevoi fotoelectrichnoi sistemi. *Informatciini tehnologii v modeluvanni: Materiali V vseukrainskoi nauk.-prakt. konf. studentiv, aspirantiv ta molodih vchenih*. Odesa. ONPU. 97-100.
18. Chaikovska, E. E., Hega K. V. (student) (2020). Matematichne obhruntuvannia pidtrimki funktcionuvannia meregevoi vitro-soniachnoi electrichnoi sistemi. *Informatciini sistemi priiniattia richen i problemi obchisluvalnoho intelectu – ISDMCI'2020*: materiali *mignar. nauk. konf.* Kherson. 154-155.
19. Chaikovska, E. E., Hega K. V. (student) (2020). Matematichne obhruntuvannia teplonasosnoho enerhopostachannia u skladi meregevoi vitro-soniachnoi elektrichnoi sistemi. *Pratci 9 Mign. nauk.-prakt. konf. "Problemi informatiki ta komputernoi tehniri" (PIKT - 2020)*, Chernivtsi, 123-125.

20. Chaikovska, E. E., Shtapenko S. V. (student) (2021). Matematichne obhruntuvannia pidtrimki funktcionuvannia fotoelectrichnoi zariadnoi stantcii. *Tezi dop. XVII Mignarodn. nauk.-prakt. konf. «Enerhetichni ta teplotechnichni protcessi i ustatcuvannia»,* Kharkiv: HTU«HPI». 90–91. ISBN 978-617-7476-56-5.
21. Chaikovska, E. E. (2022). Kompleksne upravlinnia akumuluvanniam u skladi meregevoi soniachnoi elektrichnoi sistemi. *Pratci XI Mign. nauk.-prakt. konf. "Problemi informatiki ta komputernoi tehniri" (PIKT -2022),* Chernivtsi, 112-114.

Index

A

B

C

D

E

F

H

I

M

O

P

R

S

T

W